Contributions from Breeding Forage and Turf Grasses

Contributions from Breeding Forage and Turf Grasses

Proceedings of a symposium sponsored by Divisions C-1, C-5, and C-6 of the Crop Science Society of America in Atlanta, Georgia, 1 Dec. 1987.

Editors

D. A. Sleper, K. H. Asay, and J. F. Pedersen

CSSA Special Publication Number 15

Crop Science Society of America, Inc.
Madison, Wisconsin, USA
1989

Cover Design: Patricia Jeffson

Crop Science Society of America, Inc.
677 South Segoe Road, Madison, WI 53711, USA

Library of Congress Cataloging-in-Publication Data

Printed in the United States of America.

CONTENTS

FOREWORD

Among plants, the Gramineae, or grass family, provides the major component of human foods, either directly as cereal grains or indirectly by meat animals that graze the land or are provided grasses such as hay, silage, or freshly cut forage. Native stands of grasses throughout the world have sustained wildlife and later domesticated animals. Some of the wild grasses contributed genes to important cereal grains. Others have been collected and transported to gardens for evaluation and new uses. Still others have been used in hybridization for selection of new characteristics. Thus in less than 100 yr, the world's agriculture has benefitted from genetically improved grasses. Likewise, grasses have been selected for intensely managed and heavily used sites such as sports fields and parks.

The versatility of grasses is unsurpassed among plants. This small book provides documentation for some of the advances made in genetic improvement of grasses. The smallness of the book and the relatively few scientists who have dedicated their professional careers to grass breeding severely underplay the great importance of the contributions of grass breeding to society. It does, however, provide a benchmark upon which future gains may be judged. It also establishes the advances made by traditional plant breeding practices and emphasizes the genetic complexity in a cultivar adapted to special conditions. This complexity of adaptation must be recognized as molecular-based plant breeding emerges as a reality.

The organizers of the symposium are complimented for the timeliness of their efforts. It is hoped that this book will stimulate further investments in research on grass breeding because the opportunities are great.

Calvin O. Qualset, president
Crop Science Society of America

PREFACE

What are the contributions to humankind as the result of breeding forage and turf grasses? This publication attempts to answer this question as we look back over the history of grass breeding in the United States. The contributions from forage grass breeding have resulted in significant improvements in herbage and seed yield, insect and disease resistance, herbage quality, and animal performance, to mention a few. Turf grass breeding can also claim significant advancements in the areas of adaptation of species to environmental stresses, insect and disease resistance, higher seed yields, and better turf quality for a multitude of uses.

Forage and turf grass breeders have a rich history of cooperating with scientists from other disciplines to accomplish their research objectives. Within the chapters of this publication, it is apparent that significant contributions by the forage grass breeder have come about through cooperation with agronomists, animal scientists, pasture management specialists, economists, plant physiologists, entomologists, plant pathologists, and forage chemists. Likewise, the turf breeding effort has been enhanced by receiving cooperation from horticulturalists, irrigation specialists, plant pathologists, entomologists, and others. Collectively, their efforts have resulted in forage and turf grasses being established on vast acreages in the United States—far more than for any other individual crop species.

The breeding and genetics of forage and turf grasses has not received the amount of research support that many cash crops have in the United States. Vast acreages of forage and turf grasses are grown each year which make significant contributions to the livestock industry, wildlife conservation, soil and water conservation, recreation, and have tremendous aesthetic value, yet the amount of dollars spent and the number of scientists involved is far less than what is needed. With this impetus, the Grass Breeder's Work Planning Conference appointed a committee to organize a symposium to document the contributions made by the forage and turf grass breeders. The symposium was held at the annual meetings of the Crop Science Society of America (CSSA) at Atlanta, GA in 1987. This publication is the result of that symposium.

The most important attribute of any publication is the quality of its authorship. The authors of this publication are recognized as experts in their respective scientific disciplines. A great deal of gratitude is extended to the authors for their scholarly writings and enthusiasm for the topic at hand. Thanks is also due to the CSSA for sponsoring this important publication.

Editors:

D. A. Sleper
K. H. Asay
J. F. Pedersen

CONTRIBUTORS

K. H. Asay	Research Geneticist, USDA-ARS, Forage and Range Research Laboratory, Utah State University, Logan, Utah
R. E. Barker	Research Geneticist, USDA-ARS, National Forage Seed Production Research Center, Corvallis, Oregon. Formerly with USDA-ARS, Northern Great Plains Research Laboratory, Mandan, North Dakota
Glenn W. Burton	Research Geneticist, USDA-ARS and University of Georgia, Coastal Plain Experiment Station, Tifton, Georgia
Philip Busey	Associate Professor of Ornamental Horticulture (Turfgrass Breeder), University of Florida, Fort Lauderdale, Florida
C. Reed Funk	Professor, New Jersey Agricultural Experiment Station, Soils and Crops Department, Cook College, Rutgers University, New Brunswick, New Jersey
H. J. Gorz	Supervisory Research Geneticist, USDA-ARS, Department of Agronomy, University of Nebraska, Lincoln, Nebraska
F. A. Haskins	George Holmes Professor of Agronomy, Department of Agronomy, University of Nebraska, Lincoln, Nebraska
R. R. Kalton	Forage Breeder, VISTA, Land O'Lakes Research Farm, Webster City, Iowa
Gordon C. Marten	Research Agronomist and Professor, USDA-ARS and University of Minnesota, St. Paul, Minnesota
William A. Meyer	President—Turfgrass Breeder, Pure Seed Testing, Inc., Hubbard, Oregon
D. A. Sleper	Professor of Agronomy, Department of Agronomy, University of Missouri, Columbia, Missouri
K. P. Vogel	Research Geneticist, USDA-ARS, Department of Agronomy, University of Nebraska, Lincoln, Nebraska

Conversion Factors for SI and non-SI Units

Conversion Factors for SI and non-SI Units

To convert Column 1 into Column 2, multiply by	Column 1 SI Unit	Column 2 non-SI Unit	To convert Column 2 into Column 1, multiply by
		Length	
0.621	kilometer, km (10^3 m)	mile, mi	1.609
1.094	meter, m	yard, yd	0.914
3.28	meter, m	foot, ft	0.304
1.0	micrometer, μm (10^{-6} m)	micron, μ	1.0
3.94×10^{-2}	millimeter, mm (10^{-3} m)	inch, in	25.4
10	nanometer, nm (10^{-9} m)	Angstrom, Å	0.1
		Area	
2.47	hectare, ha	acre	0.405
247	square kilometer, km^2 $(10^3\ m)^2$	acre	4.05×10^{-3}
0.386	square kilometer, km^2 $(10^3\ m)^2$	square mile, mi^2	2.590
2.47×10^{-4}	square meter, m^2	acre	4.05×10^3
10.76	square meter, m^2	square foot, ft^2	9.29×10^{-2}
1.55×10^{-3}	square millimeter, mm^2 $(10^{-6}\ m)^2$	square inch, in^2	645
		Volume	
9.73×10^{-3}	cubic meter, m^3	acre-inch	102.8
35.3	cubic meter, m^3	cubic foot, ft^3	2.83×10^{-2}
6.10×10^4	cubic meter, m^3	cubic inch, in^3	1.64×10^{-5}
2.84×10^{-2}	liter, L (10^{-3} m^3)	bushel, bu	35.24
1.057	liter, L (10^{-3} m^3)	quart (liquid), qt	0.946
3.53×10^{-2}	liter, L (10^{-3} m^3)	cubic foot, ft^3	28.3
0.265	liter, L (10^{-3} m^3)	gallon	3.78
33.78	liter, L (10^{-3} m^3)	ounce (fluid), oz	2.96×10^{-2}
2.11	liter, L (10^{-3} m^3)	pint (fluid), pt	0.473

		Mass	
2.20×10^{-3}	gram, g (10^{-3} kg)	pound, lb	454
3.52×10^{-2}	gram, g (10^{-3} kg)	ounce (avdp), oz	28.4
2.205	kilogram, kg	pound, lb	0.454
0.01	kilogram, kg	quintal (metric), q	100
1.10×10^{-3}	kilogram, kg	ton (2000 lb), ton	907
1.102	megagram, Mg (tonne)	ton (U.S.), ton	0.907
1.102	tonne, t	ton (U.S.), ton	0.907
		Yield and Rate	
0.893	kilogram per hectare, kg ha^{-1}	pound per acre, lb $acre^{-1}$	1.12
7.77×10^{-2}	kilogram per cubic meter, kg m^{-3}	pound per bushel, bu^{-1}	12.87
1.49×10^{-2}	kilogram per hectare, kg ha^{-1}	bushel per acre, 60 lb	67.19
1.59×10^{-2}	kilogram per hectare, kg ha^{-1}	bushel per acre, 56 lb	62.71
1.86×10^{-2}	kilogram per hectare, kg ha^{-1}	bushel per acre, 48 lb	53.75
0.107	liter per hectare, L ha^{-1}	gallon per acre	9.35
893	tonnes per hectare, t ha^{-1}	pound per acre, lb $acre^{-1}$	1.12×10^{-3}
893	megagram per hectare, Mg ha^{-1}	pound per acre, lb $acre^{-1}$	1.12×10^{-3}
0.446	megagram per hectare, Mg ha^{-1}	ton (2000 lb) per acre, ton $acre^{-1}$	2.24
2.24	meter per second, m s^{-1}	mile per hour	0.447
		Specific Surface	
10	square meter per kilogram, m^2 kg^{-1}	square centimeter per gram, cm^2 g^{-1}	0.1
1000	square meter per kilogram, m^2 kg^{-1}	square millimeter per gram, mm^2 g^{-1}	0.001
		Pressure	
9.90	megapascal, MPa (10^6 Pa)	atmosphere	0.101
10	megapascal, MPa (10^6 Pa)	bar	0.1
1.00	megagram per cubic meter, Mg m^{-3}	gram per cubic centimeter, g cm^{-3}	1.00
2.09×10^{-2}	pascal, Pa	pound per square foot, lb ft^{-2}	47.9
1.45×10^{-4}	pascal, Pa	pound per square inch, lb in^{-2}	6.90×10^3

continued on next page

Conversion Factors for SI and non-SI Units

To convert Column 1 into Column 2, multiply by	Column 1 SI Unit	Column 2 non-SI Unit	To convert Column 2 into Column 1, multiply by
		Temperature	
1.00 (K − 273)	Kelvin, K	Celsius, °C	1.00 (°C + 273)
(9/5 °C) + 32	Celsius, °C	Fahrenheit, °F	5/9 (°F − 32)
		Energy, Work, Quantity of Heat	
9.52×10^{-4}	joule, J	British thermal unit, Btu	1.05×10^{3}
0.239	joule, J	calorie, cal	4.19
10^{7}	joule, J	erg	10^{-7}
0.735	joule, J	foot-pound	1.36
2.387×10^{-5}	joule per square meter, J m^{-2}	calorie per square centimeter (langley)	4.19×10^{4}
10^{5}	newton, N	dyne	10^{-5}
1.43×10^{-3}	watt per square meter, W m^{-2}	calorie per square centimeter minute (irradiance), cal cm^{-2} min^{-1}	698
		Transpiration and Photosynthesis	
3.60×10^{-2}	milligram per square meter second, mg m^{-2} s^{-1}	gram per square decimeter hour, g dm^{-2} h^{-1}	27.8
5.56×10^{-3}	milligram (H_2O) per square meter second, mg m^{-2} s^{-1}	micromole (H_2O) per square centimeter second, μmol cm^{-2} s^{-1}	180
10^{-4}	milligram per square meter second, mg m^{-2} s^{-1}	milligram per square centimeter second, mg cm^{-2} s^{-1}	10^{4}
35.97	milligram per square meter second, mg m^{-2} s^{-1}	milligram per square decimeter hour, mg dm^{-2} h^{-1}	2.78×10^{-2}
		Plane Angle	
57.3	radian, rad	degrees (angle), °	1.75×10^{-2}

	Electrical Conductivity, Electricity, and Magnetism		
10	siemen per meter, S m^{-1}	millimho per centimeter, mmho cm^{-1}	0.1
10^4	tesla, T	gauss, G	10^{-4}
	Water Measurement		
9.73×10^{-3}	cubic meter, m^3	acre-inches, acre-in	102.8
9.81×10^{-3}	cubic meter per hour, $m^3\ h^{-1}$	cubic feet per second, $ft^3\ s^{-1}$	101.9
4.40	cubic meter per hour, $m^3\ h^{-1}$	U.S. gallons per minute, gal min^{-1}	0.227
8.11	hectare-meters, ha-m	acre-feet, acre-ft	0.123
97.28	hectare-meters, ha-m	acre-inches, acre-in	1.03×10^{-2}
8.1×10^{-2}	hectare-centimeters, ha-cm	acre-feet, acre-ft	12.33
	Concentrations		
1	centimole per kilogram, cmol kg^{-1} (ion exchange capacity)	milliequivalents per 100 grams, meq 100 g^{-1}	1
0.1	gram per kilogram, g kg^{-1}	percent, %	10
1	milligram per kilogram, mg kg^{-1}	parts per million, ppm	1
	Radioactivity		
2.7×10^{-11}	bequerel, Bq	curie, Ci	3.7×10^{10}
2.7×10^{-2}	bequerel per kilogram, Bq kg^{-1}	picocurie per gram, pCi g^{-1}	37
100	gray, Gy (absorbed dose)	rad, rd	0.01
100	sievert, Sv (equivalent dose)	rem (roentgen equivalent man)	0.01
	Plant Nutrient Conversion		
	Elemental	***Oxide***	
2.29	P	P_2O_5	0.437
1.20	K	K_2O	0.830
1.39	Ca	CaO	0.715
1.66	Mg	MgO	0.602

1 Contributions from Breeding Forage and Turf Grasses—An Overview

K. H. Asay
USDA-ARS
Utah State University
Logan, Utah

D. A. Sleper
University of Missouri
Columbia, Missouri

Forage grasses occupy more land area in the USA than any other crop and are of immeasurable value for livestock and wildlife feed. According to the 1986 report from the Council for Agricultural Science and Technology (CAST) (1986), forages supplied an estimated 83% of the nutrients consumed by beef cattle in the USA. Moreover, the fibrous grass root is our most effective means of soil conservation and a suitable substitute for grass in lawns, parks, athletic fields, playgrounds, golf courses, and multiple other turf and aesthetic applications has not been found. It is evident that grasses will have a more significant impact in the future. Recent government legislation provides for development of a conservation reserve to keep highly erodible land in forages and trees, and reduction of crop surpluses by promoting forages as alternatives to the crops in surplus. Increasing costs for fossil fuels and feed grains along with consumer demand for leaner meat products and less fat in the diet also will lead to more dependence on forage grasses.

Compared to other cash crops of similar economic value, grasses have received far less than their share of research support for genetics and breeding. From 1950 to 1970, most agricultural experiment stations and the USDA-ARS actively supported research projects related to breeding and genetics of forage and turf grasses. Since 1980, however, funding for this type of research activity has declined markedly. According to Kalton (1987), public grass breeding projects have been eliminated or severely curtailed in Indiana, Illinois, Minnesota, Virginia, South Dakota, Arizona, Kansas, Idaho, Washington, Oregon, New Mexico, and South Carolina. Private breeding programs, notably Northrup King and North American Plant Breeders, have also declined. The negative trend in research support for genetic improvement of perennial forage and turf grasses was confirmed by the report of the Crop Advisory Committee for Grasses (1988).

Kalton (1987) expressed concern that the decline in grass breeding research would not only lead to deficiencies in improved grass germplasms, but would be accompanied by a decline in basic research and a shortage of properly prepared grass geneticists and breeders in the future. In view of the far-reaching effects of this trend, the Grass Breeder's Work Planning Conference passed a resolution in 1984 (Table 1–1). This manifesto was sustained by several organizations including the Crop Science Society of America, Western Region Coordinating Committee WRCC-11, North Central Region Coordinating Committees NCR-36 and NCR-10, North East Regional Technical Committee NE-144, Southern Pasture and Forage Crop Improvement Conference, and the National Council of Commercial Plant Breeders. Although the negative trend in support of grass breeding is alarming, it is encouraging that several strong grass breeding and genetics programs remain intact. According to a survey that Kalton (1987), conducted breeding and enhancement programs are progressing on at least 37 different grasses, with

Table 1–1. Resolution drafted and passed by the Grass Breeder's Work Planning Conference.

WHEREAS, much of the world land area consists of forage and grasslands, including approximately 75% of the land surface of the contiguous United States. The livestock industry is largely dependent or based entirely on the grassland resource for its existence and survival.

AND WHEREAS, grasses have enhanced the agricultural base of the United States and all the Americas by allowing productive use of marginal lands and by revegetating or stabilizing lands eroded by wind and water. Abused rangelands and areas badly disturbed by man have been reclaimed and production increased with improved cultivars of grasses.

AND WHEREAS, grasses have provided for our aesthetic and recreational needs and environmental enhancement of our communities by reducing environmental pollution; and have contributed substantially to our mental, physical, and social well being.

AND WHEREAS, grass breeding, although a relatively new and unexploited research area compared with that of other major world crops, has produced many valuable cultivars for both forage and turf. Even greater progress can be expected as new technologies to transfer and fix genetic traits are developed and made available to breeders.

AND WHEREAS, the number of grass breeding programs throughout the United States, especially in forage grasses, has declined in recent years. Moreover, many of the pioneer scientists involved in grass improvement have reached or are approaching retirement. Serious concerns regarding this trend have been expressed by many professional organizations and other components of our society.

THEREFORE, BE IT RESOLVED, that this trend be reversed before irreparable damage to our agricultural productivity and aesthetic resources occurs. That we recognize the need for continued development of improved cultivars that are productive, resistant to environmental stresses and plant pests, palatable and nutritious to the grazing animal, and meet our demands for turf and soil stabilization.

WE FURTHER RESOLVE, that those involved in setting priorities for public and private research needs be encouraged to increase funding for forage and turf grass improvement programs and to replace those scientists who have retired or left their positions in recent years.

orchardgrass (*Dactylis glomerata* L.), tall fescue (*Festuca arundinacea* Schreb.), and smooth bromegrass (*Bromus inermis* Leyss.) receiving the most attention. The USDA-ARS, alone or in cooperation with State agricultural experiment stations, has taken the lead with research in many perennial grasses. This includes breeding and genetic research on 27 of the 37 grasses cited by Kalton (1987). The USDA-ARS is particularly instrumental in research programs involving grasses with limited seed volume. These include grasses adapted to semiarid rangelands of the Great Plains, the Southwest, and the Intermountain Region. In general, these grasses have regional market potentials, with limited demand for seed in any one state. State agricultural experiment stations make major contributions to breeding and genetic research involving several grasses that are important within their state boundaries. Private industry must restrict their investments to relatively high volume species that will likely return a profit from seed sales. Species such as tall fescue and orchardgrass are receiving the most attention from these programs.

The economic and aesthetic demands on America's grasslands dictates that research programs related to genetic improvement of grasses be expanded.

Recent advances in molecular biology have given the grass geneticist an exciting new arsenal of tools. Introgression of genetic traits across previously insurmountable barriers is now a realistic breeding objective. The grass geneticist, working cooperatively with other scientists representing disciplines such as molecular genetics, physiology, animal science, entomology, pathology, and biochemistry, stands on the threshold of major genetic advances. This is not the time to reduce support for these programs. To do so would not only sacrifice sizable research investments made to date, but would prevent us from exploiting new technology and unique germplasm now available to us.

This special publication is the proceedings of a symposium sponsored by the Grass Breeder's Work Planning Conference during the 1987 annual meetings of the American Society of Agronomy. Prominent scientists representing a broad array of disciplines were assembled to discuss the past accomplishments and future potential of grass breeding research involving forage and turf grasses. Excellent examples of breeding progress were presented. Several grass cultivars generated by these research programs have made and are continuing to make a major impact. In several cases, economic benefits from these improved sources of germplasm far exceeded the input costs. It is our contention that policymakers should consider past accomplishments and potential for genetic progress in the future in resource allocation for grass breeding research.

REFERENCES

Kalton, R.R. 1987. Forage grass breeding today in the U.S.A. p. 18–19. *In* Proc. 33rd Annu. Farm Seed Conf., Kansas City, MO. 3 Nov. ASTA, Washington, DC.

Council for Agricultural Science and Technology. 1986. Forages: Resources for the future. Rep. 108. CAST, Ames, IA.

Crop Advisory Committee for Grasses. 1988. Status report of crop Advisory committee for Forage and Turf Grasses, USDA-ARS. Reproduced by the Utah Agric. Exp. Stn.

2 Cool-Season Forage Grass Breeding: Progress, Potentials, and Benefits[1]

R. E. Barker

USDA-ARS
Mandan, North Dakota

R. R. Kalton

Land O'Lakes Research Farm
Webster City, Iowa

ABSTRACT

Cool-season grasses grow rapidly and initiate flowering when temperatures are moderate. Many of the world's most important forages upon which the animal industry is founded are included in this group. Two conditions are necessary to optimize grassland production: (i) proper management and (ii) use of species and cultivars adapted to the specific climate and forage purpose. Grass breeding is a relatively new science, but in the past 30 yr cultivars have been developed to provide improved feed for livestock, to conserve and improve our soil and water resources, and to provide a primary base for aesthetic and recreational purposes. Because of the diversity of uses, it is difficult, if not impossible, to measure the ultimate benefit of grass cultivars in monetary terms. Indeed, grassland agriculture, including grass cultivars, is the foundation on which life and civilization rest. Grass breeding programs have received low priority in recent years as research budgets have tightened. Unfortunately, fewer breeders are now actively pursuing improvement of the germplasm resources necessary to sustain efficient productivity of vast grassland acreages.

A wise and experienced researcher commented several years ago, "Many grasses are perennial and some are eternal." Indeed, grasses seem to be everywhere and at all times present; so much so that we tend to take them for granted. There are more species in the grass family than any other plant family. Grasses are the widest taxonomic group among plants and extend from frigid northern climates to hot, dry, southern deserts. Grasses serve

[1] Joint contribution from the USDA-ARS, Northern Great Plains Res. Lab., Mandan, ND 58554; and Vista Land O'Lakes Research Farm, Route 2, Webster City, IA 50595.

 Contributions from Breeding Forage and Turf Grasses, CSSA Special Publication no. 15.

as the primary base of the animal industry with pastures, range, haylage, greenchop, and hay contributing the feedstuffs for livestock (Hitchcock, 1951). Necessary energy for growing and maintaining animals is largely obtained from grasses that are consumed by grazing animals. Grasses also are extensively used for conservation of our natural resources from water erosion, for stabilization of blowing and drifting soils, for revegetation of land disturbed by man, for beautification of our homes and highways, and for our development of outdoor recreation activities. In the foreword of the 1948 Yearbook of Agriculture, the Honorable Clinton P. Anderson (1948) stated, "Grass can make beautiful the hillsides, schoolyards, roadsides, farmsteads; in doing so it brings greater utility and efficiency. Beauty brings serenity, and serenity is a quality we and the troubled world need." Indeed, grasses contribute beyond measure to humanity.

Cool-season grasses are those that grow in the cool moist times of the year and tend to go dormant or be injured during hot dry periods. They are found in the temperate parts of the world. Among the cool-season grasses are some of the most important crops available to humanity. The small grains such as wheat (*Triticum aestivum* L.), oat (*Avena sativa* L.), and barley (*Hordeum vulgare* L.) are included in this group of grasses. While we usually consider the small grains as crop plants, we tend to consider the other grasses in a different way. "Many people. . .look upon grassland as the foundation of security in agriculture. They believe in grass. . .the way we believe in the practice of conservation, or in good farming, or prosperity, or cooperation. For grass is all those things; it is not just a crop." (Anderson, 1948).

HISTORY OF COOL-SEASON GRASS IMPROVEMENT

Selection and improvement of grasses that have benefited people started earlier than humanity itself. Certainly, in prehistoric times as animals grazed the earth, there was natural selection among the grass species on which they fed. In biblical records, grasses and grasslands were important to the survival of humanity (Heath et al., 1985, p. xi-xii). However, grass breeding as a science is relatively young when compared with the cereals (van Bogaert, 1977). In the 1930s, grass breeding programs began when few forage researchers were trained in genetics. There were reports of "strain building" in the 1920s, but prior to that time germplasm work involved only collection, introduction, and evaluation of grass species.

In the early years of forage work, little genetic information was known about common important grasses that are currently in use as forages. Many species were evaluated for adaptation and possible forage use with the only documentation being inclusion of comments in station annual reports (Dillman, 1946). Often, collections of exotic species were made in other countries then introduced into the USA for evaluation in specific environments. In the late 1800s early 1900s, N.E. Hansen, a noted plant explorer, made collection trips into the USSR and Siberia. His original introductions of rangeland grasses were the genetic base for many cultivars still in use.

Not until the mid-1930s did breeding work on grasses progress at an appreciable rate. At that time, a few forage researchers obtained a knowledge of genetics and conducted selection among the strains and land races that had been evaluated in earlier experiments (Breese & Davies, 1969).

Early breeders of forage grasses were concerned about the same difficulties in breeding grasses that we experience today. Sources of heritable variation needed to be identified, and techniques to rapidly screen for important plant traits had to be developed. The major forage grasses have perennial growth habits that complicate accurate measurement and evaluation of forage yields. In addition, high chromosome numbers and high ploidy levels cause the genetics of most grasses to be complex (Table 2–1).

Most grasses are classified autopolyploids, allopolyploids, or autoallopolyploids. Orchardgrass (*Dactylis glomerata* L.), for example, has been assumed to be an autopolyploid. However, Lentz et al. (1983) reported both tetrasomic and disomic segregation patterns for reaction to the stem rust fungus, *Puccinia graminis* Pers f. sp. *dactylidis* Guyot et Massenot. Most of the major roadblocks to forage grass breeding are better understood today, and progress in forage improvement is advancing more rapidly with utilization of more detailed genetic knowledge. Many breeding programs are using some form of recurrent restricted phenotypic selection (Burton, 1982) to increase frequencies of favorable genes for specific traits within individual grass populations. Others are relying on outcross progeny performance for evaluation of combining ability for desired traits to select superior clonal material for synthetic recombinations. However, progress from selection for increased forage yield per se above that of original collections and introductions has been difficult or fully unsuccessful in many species (van Bogaert, 1977). Part of the problem is our inability to analyze or evaluate in a consistent manner the dynamic nature of forage dry matter accumulation. There is no specified point in time, as in cereals, when yield is finalized and can be fully evaluated. Further, coordination among researchers for cultivar evaluation has been difficult or lacking because of large environmental differences among test sites, high expense in conducting multiple site evaluations of forages, and shortages of funds. Few regional testing programs exist.

Table 2–1. Prevalent chromosome numbers and ploidy levels of common cool-season forage grasses.

Species	Chromosome no.	Ploidy
Italian ryegrass	14	Diploid
Orchardgrass	28	Tetraploid
Reed canarygrass	28	Tetraploid
Tall fescue	42	Hexaploid
Timothy	42	Hexaploid
Smooth bromegrass	56	Octoploid
Russian wildrye	14	Diploid
Crested wheatgrass	14, 28	Diploid, Tetraploid
Intermediate wheatgrass	42	Hexaploid
Western wheatgrass	56	Octoploid
Altai wildrye	84	Dodecaploid

BENEFITS TO HUMANITY

Because of the diversity of species among cool-season forage grasses, measurement of benefits has been difficult. Data are collected for the number of hectares planted for pastures, rangeland, and hay production, but often these statistics combine all forage crops together (Jewell & Johnson, 1985). The direct monetary worth of production from all forages is sometimes estimated and reported in meetings. R.E. Howarth (1986, unpublished data) stated that in monetary terms, direct gross annual value of forage from all sources ranked third in comparison with other crops at $416 million in Saskatchewan, Canada. It was behind wheat at $1935 million and canola [*Brassica napus* L. var. *napus* and *B. rapa* var. *silvestris* (Lam.) Briggs] at $492 million. Based on 1984 statistics, hectarages of grass for hay and alfalfa/grass hay mix ranked third in the Northern Great Plains at 4.78 million ha compared with 6.07 million for spring wheat and 6.07 million for summer fallow (Jewell & Johnson, 1985).

Measures of direct production and/or land area used does not represent the full importance of forages to humans. The added value of animal products, both meat and milk, resulting from higher producing or higher quality grass cultivars should also be considered. The amount of this increase is rarely measured because the direct contribution is usually unknown. To date, the best examples and documentation are available for warm-season grasses.

Another contribution difficult to measure is the impact grasses have on control of soil erosion and the resulting increase in water quality. Questions can be raised such as: What is the value of saving 10 t of topsoil by establishing grass on an eroding field? What is the value of grasses used to stabilize and beautify our roadsides and recreational areas? What is the value of grasses in wildlife plantings to provide protection and feed? Or what is the value that can be placed upon conserving our natural resources for future generations? The list of questions is extensive, but these are items that should be considered when assessing the benefits of cool-season grasses to humanity.

Since no agencies are collecting statistical data on individual grass species, certainly little information is available on the contributions of any particular cultivar. While forage grasses in general contribute immeasurably to agricultural production worldwide, few individual species or cultivars have high impact. Usually, cultivars have regional value. For this reason, there are few forage grass breeders, and they usually direct their efforts on species that are different from one another. Grass breeders will need to more carefully document economic and other benefits from new grass cultivars so that budget decision makers can better prioritize finite economic resources.

Changes in administrative policy and laws tend to affect economic worth of grasses. For example, the 1985 Food Security Act, commonly referred to as Conservation Reserve Program (CRP) in the USA, has resulted in doubling to tripling of seed prices since 1985. Some ranchers and farmers have greatly increased their net worth by using their pastures for seed production. They have made great increases in the worth of their farms. A South

Dakota rancher reportedly sold seed from 600 ha of intermediate wheatgrass [*Thinopyron intermedium* (Host) Barkworth & D.R. Dewey] pasture at an opportune time for CRP and became a millionaire.

NUMBERS OF CULTIVARS

One measure of advance attributable to forage grass breeding is the number of cultivars that have been produced. Hanson (1972) published a comprehensive list of cultivars and strains of grasses in the USA. He listed 502 entries representing 135 species of grasses. The list included 378 cultivars of which 250 were cool-season grasses and 128 were warm-season grasses. This was the last complete published list of released cultivars. However, another source indicating number of current cultivars was obtained by examining all crop cultivars, including forage, registered by the Crop Science Society of America (CSSA). Registration of crop cultivars provides an opportunity to document formal release and to report benefits and performance of individual cultivars. Since 1926, crop cultivars have been registered by CSSA (or its

Table 2–2. Crops with more than 100 cultivars registered by the Crop Science Society of America.

Crop	Cultivars	Germplasms	Parental lines
Wheat	723	286	1
Oat	317	38	26
Grass	245	64	17
Soybean	206	97	0
Barley	205	93	0
Alfalfa	149	198	3
Sorghum	125	210	169

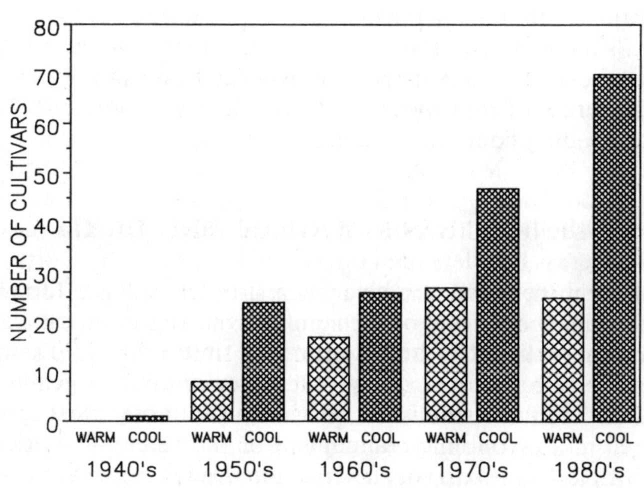

Fig. 2–1. Cool- and warm-season grass cultivars registered by the Crop Science Society of America, Madison, WI.

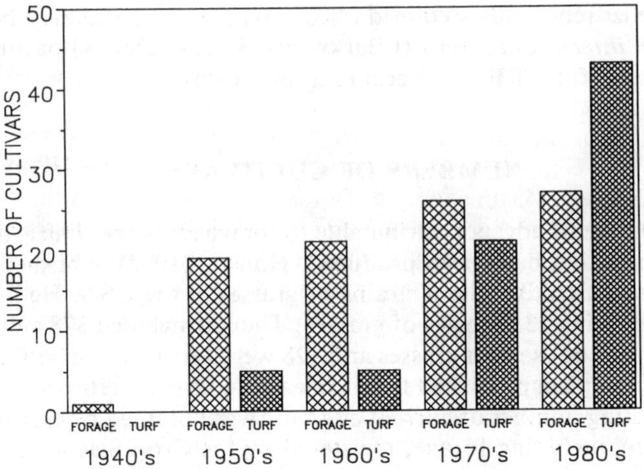

Fig. 2–2. Cool-season forage and turf grass cultivars registered by the Crop Science Society of America, Madison, WI.

parent society) (Miller & Brown, 1982). Updates to the original crop cultivar registration list are now published in the November-December issues of *Crop Science.* Wheat has the highest number of cultivars registered at 723 followed by oat at 317 (Table 2–2). Combining 12 categories of grasses together, including warm- and cool-season grasses, there have been 245 cultivars registered with CSSA (Fig. 2–1). The earliest grass cultivar was a cool-season grass registered in 1946. Since 1980, 70 cool-season grass cultivars have been registered with only 27 of the registered cultivars being forage grasses (Fig. 2–2). Forage grasses are being registered at an average rate of only two or three per year at present, while turf-type cultivars are being registered at nearly seven per year.

Rather than reenumerate the merits of each of the 94 cool-season forage grass cultivars representing 40 species registered since 1946, we will use a few selected examples of North American cultivars and use examples that demonstrate specific areas of improvement to show contributions that have been made from breeding cool-season grasses.

GENETIC BASIS OF GRASS BREEDING

Among the most significant advances assisting breeding efforts is a better understanding in recent years of the complex genetics of grasses. An example of exceptional merit is research done on the tribe Triticeae (Dewey, 1984). From extensive research over many years, Dewey defined genomes of the perennial Triticeae and established relationships among most species. The work resulted in a taxonomic realignment of the perennial Triticeae using genomic principles as the basis of classification. Dewey also made major germplasm collections representing many of the species in this important grass tribe. Grasses for use on rangelands have been improved through interspecific hybridization techniques as a result of his work (Asay & Dewey, 1985; Asay & Hansen, 1984). Wide-crossing techniques among species and genera have been extended to crossing with wheat. The total impact of this work has yet to be fully appreciated.

Elucidation of the parentage or genetic constitution of grass species may aid breeders in developing new cultivars. Western wheatgrass [*Pascopyrum smithii* (Rydb.) Löve] is an example. Dewey (1975) postulated the origin of this important rangeland grass as having its parentage from two tetraploid grasses native to ranges of western North America. They are beardless wildrye [*Leymus triticoides* (Buckl.) Pilger] and thickspike wheatgrass [*Elymus lanceolatus* (Scribn. & Smith) Gould]. Western wheatgrass is an autoallo-octoploid with four genomes. Knowledge of the genetic background allows more efficient selection and breeding techniques to be employed, but little genetic progress over original collections from the wild has been made in western wheatgrass. Perhaps parental species can first be improved and then combined to provide an improved synthetic western wheatgrass.

Collections of western wheatgrass have been made in the USA and Canada (Barker et al., 1983; Johnston et al., 1975). Considerable variation for agronomic characteristics has been identified in both populations. In the Dakota collection, variation for in vitro organic matter digestibility (IVOMD) has been observed. Among 370 individuals, IVOMD ranged from 523 to 721 g kg^{-1}. Extensive phenotypic variation was also found within families. Ranges of IVDOM from low to high individuals within a sample of 32 half-sib families varied from 11 to 136 g kg^{-1} (R.E. Barker & J.F. Karn, 1987, unpublished data). Cyclic selection has been employed for several other traits in western wheatgrass, but genetic advance within populations has been slow. Future gains from selection in this species, as well as in the parental species, will depend on the breeder's ability to use genetic information now available.

ADAPTATION TO STRESS ENVIRONMENTS

A further example of the extension of basic genetics is found in crested wheatgrass [*Agropyron desertorum* (Fisch. ex Link) Schultes]. Crested wheatgrass is a complex of diploid ($2n = 14$), tetraploid ($2n = 28$), and hexaploid ($2n = 42$) types. Tetraploids are the most prevalent among this group. Crested wheatgrass can be credited with "saving the semiarid north" in the dust bowl days in the USA because of its ability to establish and survive in drought conditions. 'Nordan' was the first cultivar to be released (Hein, 1955b). It has large seeds and compact heads with short or no awns and has been the standard for crested wheatgrass cultivars. In 1984, 'Hycrest' crested wheatgrass was released from Utah (Asay et al., 1985a). Hycrest was produced by crossing induced tetraploid *Agropyron cristatum* (L.) Gaertner s. lat. with natural tetraploid *A. desertorum*. Because of its hybrid vigor and selection pressure for establishment in drought environments, Hycrest has had about 30% higher dry matter yields than Nordan in certain stress environments (Asay et al., 1986).

Selection of cultivars for stress environments has also been demonstrated in Russian wildrye [*Psathyrostachys juncea* (Fisch.) Nevski]. Russian wildrye is well suited to grazing on western rangeland, particularly in the fall. It is a diploid grass, native to the steppe and desert regions of the USSR and People's Republic of China (PRC). 'Vinall' Russian wildrye was the first cultivar released (Hein, 1960). Vinall is a five-clone synthetic cultivar selected for high seed yield, high seed weight, lodging resistance, and high forage yield. 'Swift' Russian wildrye (Lawrence, 1980) released in Canada in 1978 is a 26-clone synthetic with selection based on ability to emerge from a 5-cm depth of seeding and coleoptile length. Because of its ability to establish, Swift had yielded 8 to 16% more dry matter forage than other Canadian cultivars (Lawrence, 1979). 'Bozoisky-Select' is a more recent cultivar released from Utah in 1984 (Asay et al., 1985b). It was developed from a plant introduction of the cv. Bozoisky. It was selected for higher seed yield, leafiness, coleoptile length, and ability to establish under drought stress. It has yielded 23% more forage than Vinall when grown in drought-stressed environments.

IMPROVEMENT IN FORAGE YIELD

Smooth bromegrass (*Bromus inermis* Leyss) is an example where selection for improved forage yield per se has been extensively practiced. Smooth bromegrass is grown widely in the USA and is known for its drought tolerance, hardiness or persistence, leafiness, and good forage production. It is a complex grass genetically, and improvement has been slow to nonexistent for forage yield. 'Lincoln' was among the first cultivars released (Hein, 1955c), and has been the standard for smooth bromegrass cultivars for many years. It was selected from a high-performing field in Nebraska that traces its origin to seed distributed prior to 1898. 'Barton' and 'Beacon' from Land O' Lakes in Iowa, were released in the early 1970s. Beacon is a Canadian-adapted cultivar and is a sister line to Barton. Both cultivars were selected for resistance to leaf spotting and leaf diseases in addition to selection for high forage yield. However, they are only 2 to 3% higher in forage yield than other cultivars (R.R. Kalton, 1987, unpublished data). 'Rebound', released in 1980, is a second-generation selection from the cv. Saratoga. Saratoga was released in 1955 with higher regrowth forage yield than Lincoln. Rebound was selected for higher regrowth ability over Saratoga, but total yields have not been any better and usually have been less than Saratoga.

Direct selection for forage yield has been difficult, and progress has been slow in all cool-season grasses. Only when selection was based upon survival or adaptation to specific stress environments such as in crested wheatgrass and Russian wildrye above, have significant increases in forage yield been shown. Forage yield increases in other grass species usually have not been significantly higher in recent cultivars than the first cultivars released.

SELECTION FOR FORAGE QUALITY COMPONENTS

While many forage grass breeding programs have concentrated on selection for higher forage yield, dry matter yield is not as important as is yield of digestible dry matter or yield of digestible energy. Each grass species has its own pattern and rate of digestible dry matter accumulation, and genetic differences exist within species. Therein lies a basic problem for selecting both for high forage yield and high forage quality (Newell & Moline, 1978). It has long been known that forage yield increases while forage quality components decrease throughout the growing season. Daugherty et al. (1982) reported polynomial equations for increase in forage yield and decrease in crude protein over time of crested wheatgrass. The basic pattern and concept is similar in other species with variability also present within a species (Fig. 2-3) (Lawrence & Warder, 1979). The genetic interaction with time emphasizes the difficulty encountered when measuring forage yield and quality. Selection for higher forage yield usually is correlated with lower quality. As expected, selection for higher quality forage has a correlated response in lower-yielding capacity (Berg & Hill, 1987). However, selecting for quality parameters without significantly altering forage yield has been reported (Casler & Ehlke, 1986; Casler, 1986).

Development of near infrared reflectance spectroscopy (NIRS) is a major breakthrough for the measurement of forage quality (Marten et al., 1985).

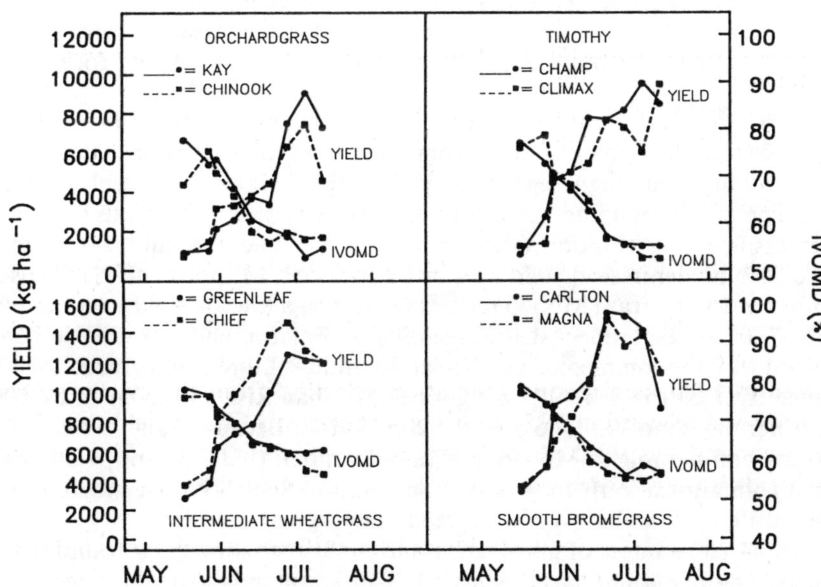

After Lawrence & Warder, 1979

Fig. 2-3. Seasonal forage yield and in vitro organic matter digestibility (IVOMD) of two cultivars in each of four cool-season grass species grown for hay in Saskatchewan, Canada.

The technology to measure many different forage quality parameters on a single sample quickly and nondestructively is of great benefit to forage breeders. Using NIRS, Ehlke et al. (1986) demonstrated that divergence between subpopulations with high and low IVDMD in smooth bromegrass became wider as the season progressed. Hence, it should be possible to select for forage quality at more growth stages and be more efficient in making genetic progress with some dynamic parameters such as cell wall constituents (Reich & Casler, 1985). It also appears probable that significant progress from selection for improved forage quality can be made while maintaining forage-yielding ability of the grasses.

SELECTION TO ELIMINATE ANTIQUALITY CONSTITUENTS

While no cool-season grass cultivars have been developed for specific forage quality parameters, several have been released that were selected for reduced antiquality characteristics. Reed canarygrass (*Phalaris arundinacea* L.) is an example. Reed canarygrass, a rugged perennial with good drought resistance, is known for its tolerance to wet soil and flooding conditions and is highly productive when managed properly. It has major problems of poor animal performance resulting from digestive disorders in cattle (*Bos taurus*) and sheep (*Ovis aries*), is highly susceptible to seed shattering, and has poor seed quality. 'Ioreed', released in 1946, was one of the first cultivars of reed canarygrass (Hanson, 1972). It was developed as being hardy, vigorous, and moderately productive, with good resistance to leaf diseases. Ioreed has a broad genetic background originating from strains of USA and foreign extraction.

'Palaton' and 'Venture' are two cultivars released in 1986. They are sister lines from the Land O'Lakes program in Iowa and were selected for ability to retain seed, high seed quality, low total alkaloid content, and no carboline and tryptamine. Carboline and tryptamine are chemicals that have been associated with poor animal performance. The new cultivars average 30 to 50% higher in seed yield over other cultivars. Marten et al. (1981) used lambs for 2 yr to graze MN-76 reed canarygrass, a low alkaloid, tryptamine-carboline-free experimental strain similar to Palaton and Venture, in comparison with the commercial cv. Rise and Vantage. Lambs that grazed MN-76 gained from 51 to 95% more weight per day than Rise and from 14 to 87% more per day than Vantage over the 2-yr period. Because Palaton and Venture are comparable to MN-76 in alkaloid content (R.R. Kalton, 1987, personal communication), they also should show superiority for grazing with beef cattle and lambs over other reed canarygrass cultivars.

Another example of eliminating antiquality constituents is found in tall fescue (*Festuca arundinacea* Schreb.). Tall fescue might be considered the queen of forage grasses because for many years more seed of this species has been sold than any of the other grasses in the USA. This grass is estimated to be grown on more than 14 million ha in the USA. Tall fescue is noted for its good summer and fall growth, but poor animal performance has been observed because of the presence of a fungal endophyte, *Acremonium coenophialum* Morgan-Jones and Gams. The endophyte is transmitted through the seed and may be eliminated by seed storage techniques.

'Alta' (Hollowell, 1945) was the first tall fescue cultivar to be bred, developed, and released. In fact, it was the first grass cultivar to be registered in the USA. Alta originated as a plant selection in 1923 and was released as a cultivar in 1940 (Hanson, 1972). As with most early cultivars, Alta was a release of an original collection or ecotype found in a field. 'Kentucky 31', another ecotype selection from a farmer's field, was released in 1942 (Fergus & Buckner, 1972). Both cultivars have been standards of comparison in tall fescue evaluation. In 1984 new cultivars, 'Mozark' and 'Martin', were released from Missouri. Martin is characterized by a higher Mg content and a broader adaptation while Mozark is more specifically adapted to the northern part of the tall fescue belt. Both cultivars are free of the fungal endophyte. The result has been approximately 30% higher average daily gains for Mozark over other cultivars in grazing trials with beef cattle (D.A. Sleper, 1987, personal communication).

'AU Triumph', a new tall fescue cultivar developed by the Alabama Experiment Station (Pedersen et al., 1983), also has shown improved animal performance. AU Triumph was developed as an early maturing cultivar to provide better distribution of forage into the fall and winter than other cultivars. In a 3-yr grazing study with steers, endophyte-free AU Triumph produced 12% more beef per hectare than endophyte-free Kentucky 31, but average daily gains were similar (Pedersen et al., 1986).

SELECTION FOR PEST RESISTANCE

Breeding grasses to tolerate or to be free of pests is an integral part of most programs. Disease resistance is the primary target since diseases affect both yield and quality. We will use orchardgrass as an example. Orchardgrass is an excellent forage grass used in pastures and for hay. It is high in forage quality and is widely adapted. A relatively large breeding effort has been concentrated on orchardgrass. 'Potomac' was released in 1954 and was the first orchardgrass cultivar registered (Hein, 1955a). Potomac originated from old pastures in the northeastern USA. Cultivars with later maturity and higher forage yields than Potomac have been released, but lower seed yields of the new cultivars resulted in poor acceptance by seed growers. 'Justus', released from Missouri in 1987, was selected specifically for resistance to stem rust (D.A. Sleper, 1987, personal communication). Stem rust has significant adverse effects on forage quality as well as a reduction of forage yield (Edwards et al., 1981). Justus was the first orchardgrass cultivar released with animal performance data. In the absence of stem rust, average daily gains over 3 yr were about equal to other cultivars. However, when stem rust was present, Justus provided 0.2 to 0.3 lb d^{-1} higher daily gains over other cultivars.

EVALUATION AND UTILIZATION OF NEW SPECIES

Finding and implementing new sources of variability or new species is a primary interest to breeders. Breeding materials that can fill a particular need are often sought. In the USA, the ryegrasses (*Lolium* spp.) have had limited use as forage grasses, while worldwide, they are probably the most widely used of all grasses. The two most important species used as forage are annual ryegrass, *L. multiflorum* Lam., and perennial ryegrass, *L. perenne* L. Breeding work on these species, particularly perennial ryegrass, has been extensive in Europe, New Zealand, and the British Isles where several excellent cultivars are available. The ryegrasses are generally considered to be higher in forage quality than such grasses as orchardgrass, bromegrass, and timothy (*Phleum pratense* L.). However, because they have not been widely adapted, only in recent years has interest from the USA been shown. Emphasis in breeding programs has been on improving annual ryegrass for tolerance to cold climates, increased rust (*Puccinia* spp.) resistance, and higher yields via induced tetraploidy. 'Urbana', a tetraploid annual ryegrass developed in Holland, has exhibited these traits when grown as a winter annual in the Deep South. Several tetraploid perennial ryegrasses are on the market for forage use. Two of these, 'Bison' and Tetrolite', have been higher yielding than common diploid perennial cultivars in many tests across the USA and have exhibited more winter hardiness, rust resistance, and seedling vigor. They appear to be well adapted for all purpose forage uses in the more southern regions where winters are mild. Phylogenetically, *Lolium* is in close proximity to *Festuca,* and numerous intergeneric hybrids have been described (Breese & Davies, 1969). 'Johnstone' tall fescue had parental clones from ryegrass × tall fescue hybrids included in its development for low perloline content (Buckner et al., 1983).

FORAGE GRASS IMPROVEMENT TEAMS

The model provided for cultivar development in recent years by reed canarygrass, orchardgrass, and tall fescue underscores the importance of animal performance evaluation in breeding forage grasses. The foremost use of a forage grass will be through the transformation of its nutrients into animal products. Without a doubt, progress has been made in forage grass breeding. We have an extensive understanding of the genetic basis for improvement. Advances have been made for forage yield and for tolerance and resistance to pests and stress environments. Progress is being made for forage quality through laboratory procedures. But animal performance is still the key to success of any forage cultivar, even though animal performance tests are expensive and time consuming.

In recent years, grass breeders have been joined by animal scientists and other disciplines to form research teams with clear objectives in mind for forage improvement. Certainly, a breeder today will probably not make significant progress alone. A team effort and cooperation from related disciplines is essential for further development and improvement of forage grass resources. We find teams of dedicated scientists wherever advances in basic understanding and development of the most promising cultivars are being made.

Forage grass breeding is in a transitional state. Past contributions can largely be attributed to individuals with a keen eye and a sense for the unusual as they selected outstanding plants from pastures and field nurseries. Most cultivars have been developed directly from ecotypic selections. Today, there is greater utilization of basic genetic understanding coupled with significant advances in instrumentation and techniques. These, in the hands of scientists from complementary disciplines, will bring new opportunities for progress in forage grass improvement.

However, austere funding is prohibiting most research teams from capitalizing on new technological developments that relate to forage grass improvement. Research programs directed toward forage grass improvement have received low resource allocations or have been eliminated over the past few years due to their apparent low economic contributions on a regional basis. Kalton conducted a survey of grass breeders in 1983 and again in 1987 (Kalton, 1983, 1987). In 1983, he asked 26 breeders about the overall support of their grass breeding programs. Only three breeders said funding support was improving while 14 said it was holding its own. Nine breeders responded that their funds were declining. In the 1987 survey, Kalton found that in recent years grass breeding programs have been lost or not replaced in about 12 states and at least two private companies. Decreases in funding were reported for 50% of the respondents. Only two respondents, one from a private company and one from a public institution, reported funding for their programs was increasing. Of the grass cultivars registered by CSSA, public institutions predominately release forage grasses while much of private industry concentrates on turfgrasses where profits may be more readily realized (Table 2–3).

Funding trends in North America seem to be following the pattern set in Europe and New Zealand (H.S. Easton, 1987, personal communication). Europe, particularly the Netherlands, has strong improvement programs in the private sector. Now that fewer tax revenues are available, New Zealand is emulating the European example. Private industry follows a market-lead system. Therefore, whatever is in highest financial demand becomes the major

Table 2–3. Grass cultivars registered by the Crop Science Society of America classified by releasing organization.

	Releasing organization		
Use	Public	Private	Mixed
Cool-season			
Forage	88	6	0
Turf	24	24	26
Warm-season			
Forage	63	1	0
Turf	9	2	2

objective of breeding programs. The immediate result seems to be breeding programs directed toward higher seed yields. Seed producers receive their main income from direct sales of seed, and they have a strong voice. However, little input is received from farmers who use grasses directly as feed in pastures. It is often difficult for farmers to see the importance of improved cultivars in their operations, because they do not have direct measures of production. Growing grasses in combination with legumes further complicates the ability to recognize the contribution from the grasses.

Funding needs to be continued at levels high enough to support team research. When funding becomes tight, teams tend to dissolve in favor of individual security. Decreased funding not only detracts from availability of improved cultivars, but also will result in less training of individuals to carry on future breeding programs. Loss of such potential talent will directly impact the strength of agriculture. Former Secretary of Agriculture, Henry A. Wallace, stated in a speech in 1940, "I have always had a great affection for grass. It seems to stand for quietness and strength. . . An agriculture without grass loses a primary source of strength." (Wallace, 1987). The statement is as true today as it was then. Grass breeding has contributed beyond measure to world agriculture. We must continue to build on this foundation. The future holds much optimistic promise for grass breeding if others are made aware of past and potential new contributions from this research.

REFERENCES

Anderson, C.P. 1948. Foreword. p. v–vi. *In* Grass—The Yearbook of Agriculture 1948. USDA. U.S. Gov. Print. Office, Washington, DC.

Asay, K.H., and D.R. Dewey. 1985. Improvement of range grasses through interspecific hybridization with quack grass [*Elytrigia repens* (L. Nevski)]. p. 256–257. *In* H. Yamaguchi (ed.) Proc. 15th Int. Grassl. Congr., Kyoto, Japan. 24–31 Aug. Natl. Grassl. Res. Inst., Nishinasuno, Tochigi-ken, Japan.

Asay, K.H., D.R. Dewey., F.B. Gomm, W.H. Horton, and K.B. Jensen. 1986. Genetic progress through hybridization of induced and natural tetraploids in crested wheatgrass. J. Range Manage. 39:261–363.

Asay, K.H., D.R. Dewey, F.B. Gomm, D.A. Johnson, and J.R. Carlson. 1985a. Registration of 'Hycrest' crested wheatgrass. Crop Sci. 25:368–369.

Asay, K.H., D.R. Dewey, F.B. Gomm, D.A. Johnson, and J.R. Carlson. 1985b. Registration of 'Bozoisky-Select' Russian wildrye. Crop Sci. 25:575–576.

Asay, K.H., and W.T. Hansen, II. 1984. Prospects for genetic improvement in the quackgrass × bluebunch wheatgrass hybrid. Crop Sci. 24:743–745.

Barker, R.E., J.D. Berdahl, J.M. Krupinsky, and E.T. Jacobson. 1983. Collections of western wheatgrass and blue grama and associated nematode genera in the western Dakotas. p. 237–240. *In* J.A. Smith and E.W. Hays (ed.) Proc. 14th Int. Grassl. Congr., Lexington, KY. 15–24 June 1981. Westview Press, Boulder, CO.

Berg, C.C., and R.R. Hill, Jr. 1987. Effect of harvest date on yield and quality of spring harvested orchardgrass. p. 56. *In* Agronomy abstracts. ASA, Madison, WI.

Breese, E.L., and W.E. Davies. 1969. Herbage plant breeding. p. 11–47. *In* Jubilee report 1919–1969. Welsh Plant Breed. Stn., Aberystwyth.

Buckner, R.C., J.A. Boling, P.B. Burrus, II, L.P. Bush, and R.A. Hemken. 1983. Registration of Johnstone tall fescue. Crop Sci. 23:399–400.

Burton, G.W. 1982. Improved recurrent restricted phenotypic selection increases bahiagrass forage yields. Crop Sci. 22:1058–1061.

Casler, M.D. 1986. Causal effects among forage yield and quality measures of smooth bromegrass. Can. J. Plant Sci. 66:591–600.

Casler, M.D., and N.J. Ehlke. 1986. Forage yield and yield component response to divergent selection for *in vitro* dry matter digestibility in smooth bromegrass. Crop Sci. 26:478–481.

Daugherty, D.A., C.M. Britton, and H.A. Turner. 1982. Grazing management of crested wheatgrass range for yearling steers. J. Range Manage. 35:347–350.

Dewey, D.R. 1975. The origin of *Agropyron smithii.* Am. J. Bot. 62:524–530.

Dewey, D.R. 1984. The genomic system of classification as a guide to intergeneric hybridization with the perennial Triticeae. p. 209–279. *In* J.P. Gustafson (ed.) Gene manipulation in plant improvement. Plenum Publ., New York.

Dillman, A.C. 1946. The beginnings of crested wheatgrass in North America. J. Am. Soc. Agron. 38:237–250.

Edwards, M.T., D.A. Sleper, and W.Q. Loegering. 1981. Histology of healthy and diseased orchardgrass leaves subjected to digestion in rumen fluid. Crop Sci. 21:341–343.

Ehlke, N.J., M.D. Casler, P.N. Drolsom, and J.S. Shenk. 1986. Divergent selection for *In vitro* dry matter digestibility in smooth bromegrass. Crop Sci. 26:1123–1126.

Fergus, E.N., and R.C. Buckner. 1972. Registration of Kentucky 31 tall fescue. Crop Sci. 12:714.

Hanson, A.A. 1972. Grass varieties in the United States. USDA Agric. Handb. 170. U.S. Gov. Print. Office, Washington, DC.

Heath, M.E., R.F Barnes, and D.S. Metcalfe (ed.) 1985. Forages—The science of grassland agriculture. 4th ed. Iowa State Univ. Press, Ames.

Hein, M.A. 1955a. Registration of varieties and strains of orchardgrass (*Dactylis glomerata*). Potomac orchardgrass. Agron. J. 47:539.

Hein, M.A. 1955b. Registration of varieties and strains of wheatgrass. II. (*Agropyron* spp.). Nordan crested wheatgrass. Agron. J. 47:546.

Hein, M.A. 1955c. Registration of varieties and strains of bromegrass (*Bromus* spp.), II. Lincoln smooth bromegrass. Agron. J. 47:533–534.

Hein, M.A. 1960. Registration of varieties and strains of other grasses, IV. Vinall Russian wildrye. Agron. J. 52:662.

Hitchcock, A.S. 1951. Introduction. p. 1. *In* Manual of the grasses of the United States. USDA Misc. Publ. 200. 2nd ed. revised by Agnes Chase. U.S. Gov. Print. Office, Washington, DC.

Hollowell, E.A. 1945. Registration of varieties and strains of grasses, I. Alta fescue, Reg. no. 1. Agron. J. 37:653–654.

Jewell, L.D., and E. Johnson (ed.) 1985. Agricultural statistics 1985. U.S. Gov. Print. Office, Washington, DC.

Johnston, A., S. Smoliak, and M.D. MacDonald. 1975. Agronomic variability in *Agropyron smithii.* Can. J. Plant Sci. 55:101–106.

Kalton, R.R. 1983. Whither grass breeding? Seed World 121(8):14–18.

Kalton, R.R. 1987. Forage grass breeding today in the U.S.A. p. 18–29. *In* Proc. 33rd. Annu. Farm Seed Conf., Kansas City, MO. 3 Nov. ASTA, Washington, DC.

Lawrence, T. 1979. Swift, Russian wild ryegrass. Can. J. Plant Sci. 59:515–518.

Lawrence, T. 1980. Registration of Swift Russian wild ryegrass. Crop Sci. 20:672.

Lawrence, T., and F.G. Warder. 1979. Evaluation of twenty grass populations as irrigated hay crops for southwestern Saskatchewan. Can. J. Plant Sci. 59:691–700.

Lentz, E.M., C.F. Crane, D.A. Sleper, and W.Q. Loegering. 1983. As assessment of preferential chromosome pairing at meiosis in *Dactylis glomerata.* Can. J. Genet. Cytol. 25:222–232.

Marten, G.C., R.M. Jordan, and A.W. Hovin. 1981. Improved lamb performance associated with breeding for alkaloid reduction in reed canarygrass. Crop Sci. 21:295–298.

Marten, G.C., J.S. Shenk, and F.E. Barton II (ed.) 1985. Near infrared reflectance spectroscopy (NIRS): analysis of forage quality. USDA Agric. Handb. 643. U.S. Gov. Print. Office, Washington, DC.

Miller, P.A., and W.L. Brown. 1982. Registered field crop varieties: 1926–1981. CSSA, Madison, WI.

Newell, L.C., and W.J. Moline. 1978. Forage quality evaluations of twelve grasses in relation to season for grazing. Univ. Nebraska Res. Bull. 283, Lincoln.

Pedersen, J.F., R.L. Haaland, C.S. Hoveland, C.D. Berry, S.P. Schmidt, and R.R. Harris. 1983. Registration of AU Triumph tall fescue. Crop Sci. 23:182.

Pedersen, J.F., J.A. McGuire, S.P. Schmidt, C.C. King, Jr., C.S. Hoveland, and L.A. Smith. 1986. Steer performance as affected by tall fescue cultivar and level of *Acremonium coenophialum* infection. N. Z. J. Exp. Agric. 14:307–312.

Reich, J.M., and M.D. Casler. 1985. Effect of maturity and alfalfa competition on expected selection response for smooth bromegrass forage quality traits. Crop Sci. 25:635–640.

van Bogaert, G. 1977. Scope for improving the yield of grasses and legumes by breeding and selection. p. 29–35. *In* B. Gilsehan (ed.) Proc. of the Intl. Meet. on Animal Production from Temperate Grassland, Dublin, Ireland. June. Gov. Plant Breed. Stn., Merelbeke, Belgium.

Wallace, H.A. 1987. Speech delivered in June 1940 during the National Broadcasting Company's National Farm and Home Hour. Reprinted in the Des Moines Register, May 6, 1987. Des Moines, IA.

3 Progress and Benefits to Humanity from Breeding Warm-Season Forage Grasses

Glenn W. Burton

USDA-ARS
University of Georgia
Tifton, Georgia

ABSTRACT

Breeders in the Western region have produced 75 cultivars of warm-season forage grasses; all but 16 are native grasses. Breeders in the Southern region have developed 50 cultivars of 10 adapted warm-season forage grass introductions. Generally, adaptation and yield have been increased, sometimes as much as 100%. Such yield increases with the same inputs greatly improve the efficiency of the livestock industry. The benefits to humanity are difficult to determine because no state or federal agricultural statistics agency collects information on the hectarage planted to warm-season grass cultivars. Cattlemen who have used the improved cultivars have benefited. One example is a Texas cattleman with 1600 ha of Coastal bermudagrass who said "We'd be out of the cattle business without Coastal." Progress to date indicates that well-supported warm-season forage grass breeding projects in which the breeder obtains the cooperation of animal science and other needed disciplines should be able to increase the yield and efficiency of most warm-season grasses 25 to 50%. Such increases without added inputs will reduce feed costs, increase the cattleman's profits, and lower the prices the consumer must pay for animal products.

Warm-season is a broad term that includes many grass species which grow best during the warm summer months, either native to the USA or introduced from foreign countries. Most warm-season grasses are restricted to the southern half of the USA because they are less winter hardy than cool-season grasses. However, cold-tolerant ecotypes occur in many warm-season species. Many warm-season forage grasses fix solar energy by the C_4 process, whereas most cool-season grasses use the less-efficient C_3 process.

Warm-season forage grasses supply well over half of the feed consumed by cattle (*Bos taurus*) in the USA. In the 1930s breeding efforts began to improve the most promising species. The few young geneticists who became the early grass breeders had first to decide which of several species adapted

 Contributions from Breeding Forage and Turf Grasses, CSSA Special Publication no. 15.

to their state or states should receive attention. Usually, they worked with several species about which little was known. Chromosome counts were needed. Flowering habits had to be determined. Mode of reproduction required study that finally explained why some species such as tetraploid bahiagrass (*Paspalum notatum* Flügge) failed to respond to selection. All bahiagrass introductions were found to be apomictic and would not release their inherent diversity until they could be crossed with a sexual female plant. Finding or creating such tetraploid females by doubling the chromosome number of sexual diploids was required before progress could be made (Burton & Forbes, 1960). Most of the promising grass species produced low seed yields. Solving the many problems associated with seed increase, distribution, and on-farm use of new cultivars required much of the breeder's time and resources. Rarely was there an opportunity to evaluate new cultivars in grazing and feeding trials with cattle. Cooperation with animal science was usually lacking and too often support was inadequate. Warm-season grass breeders sought greener pastures. Administrators, who did not realize that warm-season grass breeding was long-term and costly, closed out their research projects or assigned the breeder other tasks that left little time for grass breeding.

In light of this history of warm-season grass breeding, it is significant that Voight and MacLaughlan were able to list in 1985, 24 western warm-season grass species for which there were 75 available cultivars (Voight & McLaughlan, 1985). Fifty cultivars of 10 additional warm-season species developed in the South can be added to this list of western grasses. Progress has been made.

Warm-season forage grasses can logically be divided into two groups, Western and Southern, based largely on soil and climatic differences in the two regions. Evolutionary factors have developed native grasses in the South so poor in yield potential and quality that all pasture improvement has had to depend on introduced grass species. In the West, a number of the native grasses were good enough to warrant efforts to improve them by breeding. Of the 75 cultivars developed in western warm-season grasses, only 16 were introductions (Bashaw, 1980; Jacobson et al., 1985; Voight & Oaks, 1985). These were increased and released after their superiority over other grasses had been demonstrated in small plot tests. Most of the 16 were apomictic. Only two of the apomictic cv. Neuces and Lano buffelgrass (*Cenchrus ciliaris* L.), were the product of additional grass breeding (Bashaw, 1980). A sexual female plant crossed with an apomictic male released the variability that made it possible to increase winter hardiness, and yield by 25 to 35% in these two cultivars. Of the 59 remaining native grass cultivars, 32 were U.S. ecotypes that had been screened in small plot nurseries for various agronomic traits, often tested in similar nurseries in other environments, and then named and released as cultivars (Heitzer & Hassell, 1985; Jacobson et al., 1985; Vogel et al., 1984). The USDA-SCS Plant Materials Centers were actively involved in at least half of these developments. The other 27 native grass cultivars were developed by the USDA-ARS and State experiment station grass breeders who usually started with collections of several superior ecotypes adapted to a specific temperature-photoperiod zone.

As early as 1944, Olmstad noted that genetically, different types of *Bouteloua* occur at different latitudes (Olmsted, 1944). Later, Newell and others observed that seed of native grasses which moved too far north or south of their origin were often unadapted (Newell, 1968a). Most warm-season grasses grow taller, yield more, and flower later as they are moved north, but cannot be moved more than 402.25 to 643.6 km (250–400 mi) without experiencing winterkill (Vogel et al., 1985). Northern ecotypes moved south are shorter, flower earlier, and yield less. Photoperiod and temperature differences are responsible for these responses.

An example of cultivar development in Western warm-season grasses is the popular 'Pathfinder' switchgrass [*Panicum virgatum* (L.)]. In a collection of adapted ecotypes, Newell selected "f"-type clones, polycrossed them, and tested their progenies (Newell, 1968b). Twelve clones producing the best progenies were polycrossed in isolation. From their progeny, 192 plants were selected and intercrossed to produce breeder seed of Pathfinder switchgrass. Modifications of this population improvement procedure were employed by USDA and State grass breeders to develop the other 26 warm-season cultivars (Anderson et al., 1989; Burton & Monson, 1988; Jacobson et al., 1985; Vogel et al., 1984).

Early warm-season grass breeders such as L.C. Newell and Jim Ross developed the initial cultivars for each temperature-photoperiod zone (K.P. Vogel, 1986, personal communication). Usually, no more than one or two cultivars of each species were developed for each zone. Rarely were they tested in feeding or grazing trials by cattle. As a consequence, my questionnaire asking for advantages over the checks and yields of forage and animal product in percent of the checks were returned with no information. Those early cultivars are the checks for the cultivars being developed today.

'Trailblazer' switchgrass is an excellent example of the development of a second-generation cultivar. Using recurrent restricted phenotypic selection in a population of switchgrass similar to Pathfinder, Vogel et al. were able to increase in one cycle, the yield of dry matter and the digestibility of the forage (IVDMD) 4.0 and 6.0%, respectively, when compared with Pathfinder for 3 yr (Vogel et al., 1985). In a 3-yr replicated grazing trial, Trailblazer gave 23% better average daily gains than Pathfinder and produced 23% more liveweight gain per 0.4 ha.

All warm-season grasses used in southern-improved pastures originated in foreign countries. Of the many species that have been tested, only 10 are being used or are likely to be used in the near future. No cultivars of carpetgrass (*Axonopus affinis* Chase) nor johnsongrass [*Sorghum halapense* (L.) Pers.] (considered one of the 10 worst weeds in the world) have been developed.

Three of the species in use or of interest today are obligate apomicts. The 50-chromosome dallisgrass (*Paspalum dilatatum* Poir.) has been the subject of much study and improvement effort. None of numerous introductions from its native South America significantly improved the forage properties or the poor seeding habit of common dallisgrass. Interspecific hybridization failed and natural or induced mutants were no better than the apomictic common dallisgrass. As a result, no U.S. cultivars of common dallisgrass have been developed and named.

The tetraploid bahiagrasses that occupy millions of hectares in South America are obligate apomicts, and are well adapted to the southeastern USA. Numerous introductions have been observed in nurseries and several of the better ones have been named. 'Argentine', one of several selected, increased, named, and released by the USDA-SCS Plant Materials Centers is widely grown in southern Florida. Most of the tetraploid forms lack the winter hardiness needed to be grown successfully north or Florida.

To improve the apomictic bahiagrasses, sexual tetraploids were developed by doubling the chromosomes in the sexual diploid Pensacola bahiagrass (Burton & Forbes, 1960). These sexual tetraploids used as females were crossed with apomictic males to produce highly variable progenies containing both-sexual and apomictic plants. From further breeding efforts came the apomictic 'Tifton 54' cv., superior to apomictic introductions in yield and winter hardiness, and two sexual and apomictic populations being improved by recurrent restricted phenotypic selection (RRPS).

'Pensacola' bahiagrass is the most widely grown, seed-propagated, perennial warm-season grass in the South. Increased from a few plants growing near Pensacola, FL by the USDA-SCS Plant Materials Center and named Pensacola, we now know that it arrived in the digestive tracts of cattle shipped from Santa Fe, Argentina many years ago. Compared with the apomictic tetraploid bahiagrasses, the sexual diploid Pensacola yields more forage and seed, is more winter hardy and more widely adapted, and is also easier to establish from seed.

Early attempts to improve Pensacola bahiagrass involved a search for self-incompatable cross-compatable clones that would give higher-yielding hybrids when crossed. Two such self-incompatability hybrids, 'Tifhi-1' and 'Tifhi-2', that yielded 25% more forage in clipped plots and 17% more beef gain per acre when grazed were developed. Seed was produced by harvesting all seed from seed fields planted vegetatively to alternate strips of two selected parent clones. Once established, seed fields could produce hybrid seed for many years. However, the high cost of hand labor required to establish these seed fields kept these hybrids from becoming important economically.

Improving the yield of Pensacola bahiagrass by mass selection was begun in 1960. The initial plan involved intermating the highest-yielding 200 plants in a space-planted population of 1000 plants in each cycle. Three years were required for the first cycle. Modifications in procedures to improve the precision of the screen for yield and the efficiency of the intermating system, now permit one cycle per year with the same 16% increase in space-plant yield per cycle (Burton, 1982). The eight modifications of mass selection called *restrictions* in the breeding procedure called *recurrent restricted phenotypic selection* (RRPS) have in nine cycles produced 'Tifton 9', released in 1987. Compared with Pensacola bahiagrass, Tifton 9 is much more vigorous in the seedling stage, easier to establish, and equal in forage quality (Burton, 1986). It has also yielded 47% more forage in a 3-yr seeded plot clipping test and produced up to three times more seed.

Common bermudagrass [*Cynodon dactylon* (L.) Pers.] is the most widely distributed warm-season pasture grass in the South. Originating in Africa, it probably came to America as seed in hay that the first Spaniards brought to feed their horses (*Equus caballus*). By the 1930s, common bermudagrass was a serious weed in most cultivated fields, brought there no doubt in the digestive tracts of the animals used to till the soil. Most of the southern pastures containing common bermudagrass were once cultivated fields infested with common bermudagrass.

In 1936, breeding bermudagrass began as a cooperative USDA, Georgia Coastal Plain Exp. Stn. project at Tifton. The first cv. named Coastal for the Coastal Plain Exp. Stn. was the best of 5000 F_1 hybrids between a unique common bermudagrass and two excellent introductions from South Africa (Burton, 1947). In plot tests, Coastal yielded twice as much as common and was superior to either parent, but failed to produce seed. Released in 1943, Coastal was slowly accepted due to prejudice and the need for vegetative propagation. Compared with common bermudagrass in repeated tests across the South, Coastal was more disease resistant, grew tall enough to cut for hay, yielded twice as much in good years and six times as much during the severe drought in 1954. Coastal bermudagrass used fertilizer more efficiently than other grasses. Fertilized with 330 kg/ha of N plus adequate amounts of other nutrients and cut every 6 wk, Coastal consistently produced 22 t of hay per hectare. In grazing trials, Coastal bermudagrass was more palatable and produced twice as much beef as common (Burton, 1947).

Since Coastal's release and widespread use across the South, 11 other vegetatively propagated hybrids have been developed and used. The first hybrid was 'Midland', a more winter-hardy hybrid produced by crossing Coastal with a winter-hardy common bermudagrass from Indiana. 'Hardie', developed in Oklahoma, and 'Tifton 44' (the best of some 3500 F_1 hybrids between Coastal bermudagrass and a winter-hardy bermuda from Berlin, Germany) are others widely grown north of the Coastal bermudagrass belt.

Crossing Coastal with a highly digestible (in vitro dry matter digestibility, IVDMD) but cold susceptible introduction from Kenya produced 'Coastcross-1' (the best of many) that yielded as much dry matter and was 12% more digestible than Coastal (Burton, 1972). In feeding and grazing trials, Coastcross-1 produced 30 to 40% better average daily gains (ADGs) and liveweight gains per hectare (LWG/ha) than Coastal. Coastcross-1 lacks winter hardiness but is widely used in Florida and tropical countries. In Cuba, it occupies more than 230 000 ha and produces 4500 kg of milk per lactation without concentrate.

'Tifton 78' bermudagrass, the most recently released hybrid, is the best of many F_1 hybrids between Tifton 44 and the cold and rust susceptible but vigorous 'Callie'. Tifton 78 carries the rust resistance, drought and cold tolerance of Coastal, and the excellent vigor and ease of establishment of Callie. Compared with Coastal in a 3-yr replicated grazing trial, Tifton 78 produced 15% better ADGs and 36% more LWG/h (Burton & Monson, 1988). In this grazing trial, Tifton 78 fertilized with 168 kg/ha per year of N + P and K, carried over seven 258 kg steers/ha from 4/10 to 10/9 and produced 985 kg/ha per year of liveweight gain. Total fertilizer cost per kilogram of gain-was $0.105. Nitrogen cost per kilogram of gain was $0.074.

One of the most widely grown grasses in southern Florida and the tropics is 'Pangola grass' (*Digitaria decumbens* Stent) introduced from South Africa in 1935. Pangola grass is completely sterile and all attempts to improve it by hybridization have failed. It is established by planting freshly mowed tops and stolons. It produces no rhizomes. 'Transvala', 'Taiwan', 'Survenola', and 'Slenderstem' are vegetatively propagated cultivars developed from other *Digitaria* spp.

'Carostan' flaccidgrass (*Pennisetum flaccidum* Griseb.) is a newly released apomictic cultivar developed by D.H. Timothy, North Carolina State Univ., Raleigh from PI 220606 from Afghanistan (D.H. Timothy, 1987, personal communication). Although it has excellent quality and yield, its poor seeding characteristics may limit its use.

Napiergrass (*Pennisetum purpureum* Schumach), introduced from Africa in 1913, is a robust perennial bunch grass that resembles sugarcane (*Saccharum officinarum* L.) and may grow to heights of 6 m. It is grown in tropical and subtropical regions of the world where it is cut and fed to cattle. Cut once a year, it can exceed all other grasses in biomass production (Burton, 1986). Napiergrass breeding began at Tifton, GA in 1936 with the collection of seed from African introductions highly resistant to eyespot (caused by *Helminthosporium ocallum* Faris, a synonym of *Bipolaris sacchari* (Van Bredade Haan) Shoemaker—Eye spot (Florida Ring Spot). A very leafy dwarf, no. 208 with a maximum height of 1.3 m appeared in 1937 in 1000 spaced plants from this seed. Compared with tall selections, 208 yielded 9% more leaves but less than half as much total forage. Increased and grazed, 208 showed promise as a pasture grass if managed properly. In 1944, a 1.3-ha pasture of napiergrass 208 was grazed with 12 cows beginning 15 September for 25 d and produced 300 cow days of grazing plus 27 kg of LWG. In 1942, 160 F_1 hybrids between 208 and an outstanding selection, no. 1, were leafier and yielded 35% more than the best common napiergrass checks.

In the fall of 1941, napiergrass selection no. 1 was crossed with two different S-2 lines of late-maturing pearl millet. Several of these F_1 hybrids yielded more than the napiergrass checks in clipping trials (Burton, 1944).

In 1943, the best 208 × no. 1 F_1 hybrid clone and four other superior clones including two no. 1 × pearl millet hybrids were sent to experiment stations in Puerto Rico, Cuba, and Panama. Several years later, J.W. Fortuno, Experiment Station, Rio Piedras, Puerto Rico, reported that hybrid 208 × no. 1 in tests outyielded all other napiergrasses, produced 172 t/ha of green forage, and had been released under the name 'Merkeron'.

A clone of Merkeron has survived in the field at Tifton to date. When Wayne Hanna wanted to develop a dwarf napiergrass, he was able to select it in a seedling progeny of the Merkeron clone. Merkeron had preserved the dwarf gene that produced 208 for 40 yr. Since 1974, Hanna has been developing F_1 pearl millet × napiergrass hybrids, seed of which has been produced on dwarf male-sterile pearl millet planted between rows of napiergrass that serve as the pollinator (Heitzer & Hassell, 1985). Pastures planted to this seed have given excellent animal performance and have supplied grazing later in the season than pearl millet.

In 1936, sudangrass [*Sorghum* × *drummondii* (Steudel)] breeding began at Tifton with the crossing of the best sudangrasses with the disease-resistant 'Leoti' sorghum [*Sorghum bicolor* (L.) Moench]. In some 35 000 F_2 of these crosses, six plants that were about as disease resistant as Leoti but with finer stems were found (Burton, 1943). These plants were crossed with selected sudangrass plants to produce an F_2 population of some 30 000 plants, one of which was fine stemmed and as disease resistant as Leoti sorghum. A selfed progeny of this F_2 plant was as disease resistant as its parent and possessed similar plant characteristics. It was increased in isolation, named 'Tift' and released.

'Georgia 337' sudangrass combining sweet juicy stalk, low hydrogen cyanide glucoside (HCN) content, resistance to three diseases, and uniform brown seeds in straw-colored glumes was developed in four cycles from 'Tift', 'Wisconsin low HCN', 'Texas sweet', and 'McLean' sorghum (Burton, 1964). Each cycle involved intermating selected F_3s from F_2 populations ranging from 20 000 to 60 000 plants. Georgia 337 sudangrass has been used as a breeding line since higher-yielding commercial sorghum sudangrass hybrids have replaced it on the farm. Hanna is developing leafier sorghum sudangrass hybrids that should give improved animal performance when grazed, using dwarfing genes to shorten Georgia 337.

Pearl millet [*Pennisetum glaucum* (L.) R. Br.] breeding for improved forage production began at Tifton in 1936 with the development of inbred lines in the early maturing cattail millet grown at the time (Burton, 1983). Later, germplasm from India and Africa were involved. 'Starr', the first cultivar released, was produced by intermating selected F_3 progenies of a single-cross that carried the increased leafiness and later maturity desired. In clipping tests, it yielded about 25% more forage than cattail millet, the check. A second synthetic cultivar, 'Tiflate' was developed by mixing together seeds from 54 short-day introductions from Nigeria and Upper Volta. Tiflate's failure to mature seed before frost in the continental USA has restricted its usefulness to such tropical countries as Mexico and Brazil.

'Gahi-1', yielding 50% more than cattail millet, was developed by producing four inbred lines that flowered at the same time and gave high F_1 yields in all combinations. Gahi-1 seed containing 75% of hybrid seed and 25% of selfs and sibs of the four parents was produced by harvesting all seed from fields planted to a mixture of equal quantities of pure live seeds of the four parent inbreds. The more vigorous hybrid seedlings crowded out the selfs and sibs early, enabling Gahi-1 to yield as well as the double cross involving the four parents.

The discovery of cytoplasmic male sterility in pearl millet in 1954 and the development of 'Tift $23A_1$' and its maintainer 'Tift $23B_1$' permitted the production of single crosses yielding at least 50% more forage than cattail millet. Introducing the recessive d_2 gene to reduce the height of Tift A_1 and B_1 50% produced 'Tift' $23DA_1$, and 'Tift $23DB_1$' that facilitates seed production. 'Gahi-3' ('Tift $23D_1A_1$ × 'Tift 186') tops most forage yield trials. Because Tift 186 lacks the dominant *R* gene for fertility restoration, Gahi 3 sets no seed that keeps it vegetative longer and reduces its weed potential. 'Tifleaf 1' (Tift $23DA_1$ × 'Tift 383' carrying the d_2 gene) is only half as tall as Gahi 3, yields less dry matter but more leaves and gives better ADGs and LWGs/ha. 'Tift $85D_2A_1$' and 'Tift $85D_2B_1$' are similar to Tift $23DA_1$ and Tift $23DB_1$, but they carry the dominant genes for immunity to rust (*Puccinia substriata* var. *Indica*) and leafspot [*Pyricularia grisea* (Che.) Sacc.] introduced from *Pennisetum glaucum* (L.) R. Br. spp. *monodii* (Maire) Brunken (Hanna & Monson, 1980). 'Tifleaf II' (Tift $85D_2A_1$ × Tift 383) is similar to Tifleaf 1 but is immune to rust and leafspot. Commercial pearl millet F_1 hybrids are produced on Tift $23DA_1$ or other dwarf lines in A_1 cytoplasm.

Progress has been made in the USA, breeding 34 species of warm-season grasses. In the past 50 yr, plant breeders with the help of workers from other disciplines have developed more than 125 cultivars. Many of them possessing superior agronomic traits and released after limited testing are the checks for new cultivars being developed today. Generally, adaptation and yield have been increased, sometimes as much as 100%.

The benefits to humanity are more difficult to determine. Those cattlemen who have used the improved cultivars have reaped benefits similar to those described above. An example is a Texas cattleman with 1600 ha of Coastal bermudagrass who said, "We'd be out of the cattle business if we didn't have Coastal." Unfortunately, no State or Federal agricultural statistics agency collects information on the hectarage planted to warm-season grass cultivars. A number of years ago, the South Carolina extension pasture specialist, after a careful survey reported that his state had planted a million acres to Coastal bermudagrass. Estimates from other southern states at that time suggested that some 4 million ha had been planted in the South. More have been planted since then.

Certainly the few examples described above indicate that the warm-season grasses upon which the livestock industry depends can be greatly improved by breeding. Well-supported grass breeding projects, in which the breeder obtains the cooperation of animal science and all other needed disciplines, should be able to increase the yield and efficiency of most warm-season grasses 25 to 50%. Such increases with no added inputs can greatly reduce feed costs and increase profits for the cattlemen who use the improved cultivars. The consumers, such as the Cuban children who should get cheaper milk because it is produced with Coastcross-1 bermudagrass, should also benefit.

REFERENCES

Anderson, B.E., J.K. Ward, K.P. Vogel, M.G. Ward, H.J. Gorz, and F.A. Haskins. 1989. Studies with switchgrass strains selected for differing digestibility. I. Forage quality and performance of grazing yearlings. J. Anim. Sci. (in press).

Bashaw, E.C. 1980. Registration of Neuces and Llano buffelgrass. Crop Sci. 20:112.

Burton, G.W. 1943. Tift Sudan. Georgia Coastal Plain Exp. Stn. Cir. 11, April.

Burton, G.W. 1944. Hybrids between napiergrass and cattail millet. J. Hered. 35:226–232.

Burton, G.W. 1947. Breeding bermudagrass for the South-eastern United States. J. Am. Soc. Agron. 39:551–569.

Burton, G.W. 1964. Georgia 337 sudangrass. Crop Sci. 4:666.

Burton, G.W. 1972. Registration of Coastcross-1 bermudagrass. Crop Sci. 12:125.

Burton, G.W. 1982. Improved recurrent restricted phenotypic selection increases bahiagrass forage yields. Crop Sci. 22:1058–1061.

Burton, G.W. 1983. Breeding pearl millet. p. 162–182. *In* Jules Janick (ed.) Plant breeding reviews. Vol. 1. AVI Publ. Co., Westport, CT.

Burton, G.W. 1986. Tifton 9 Pensacola bahiagrass—a product of recurrent restricted phenotypic selection. p. 1. *In* Abstract of technical papers, Orlando, FL. 2–5 Feb. Southern Assoc. of Agric. Scientists and Southern Branch ASA. ASA, Madison, WI.

Burton, G.W. 1986. Biomass production from herbaceous plants. p. 163–171. *In* W.H. Smith (ed.) Biomass energy development. Plenum Press, New York.

Burton, G.W. and I. Forbes, Jr. 1960. The genetics and manipulation of obligate apomixis in common bahiagrass (*Paspalum notatum*). p. 66–71. *In* Proc. 8th Int. Grassl. Congr., Reading, Berkshire, England.

Burton, G.W. and W.G. Monson. 1988. Registration of Tifton 78 bermudagrass. Crop Sci. 28:187–188.

Hanna, W.W. and W.G. Monson. 1980. Yield, quality, and breeding behavior of pearl millet × napiergrass interspecific hybrids. Agron. J. 72:358–360.

Hanna, W.W., H.D. Wells, and G.W. Burton. 1987. Registration of pearl millet inbred parental lines, Tift $85D_2A_1$ and Tift $85D_2B_1$. Crop Sci. 27:1324–1325.

Heitzer, R.B., and W.G. Hassell. 1985. Improvement of the Gramas and other shortgrass prairie species. p. 171–177. *In* J.R. Carlson and E.D. McArthur (chair) Range plant improvement in western North America. Proc. Symp. Annu. Meet., Salt Lake City, UT. 14 Feb. Soc. of Range Manage., Salt Lake City.

Jacobson, E.T., C.M. Talilaferro, C.L. Dewald, D.A. Tober, and R.J. Haas. 1985. New and old world bluestems. p. 148–158. *In* J.R. Carlson and E.D. McArthur (chair) Range plant improvement in western North America. Proc. of Symp., Annu. Meet., Salt Lake City, UT. 14 Feb. Soc. of Range Manage. Salt Lake City.

Newell, L.C. 1968a. Effects of strain source and management practice on forage yields of two warm-season prairie grasses. Crop Sci. 8:205–210.

Newell, L.C. 1968b. Registration of Pathfinder switchgrass. Crop Sci. 8:516.

Olmsted, C.E. 1944. Growth and development in range grasses. Bot. Gaz. 106:382–401.

Vogel, K.P., K. Britton, H.J. Gorz, and F.A. Haskins. 1984. *In vitro* and *in vivo* analysis of hays of switchgrass strains selected for high and low *in vitro* dry matter digestibility. Crop Sci. 24:977–980.

Vogel, K.P., C.L. Dewald, H.J. Gorz, and F.A. Haskins. 1985. Improvement of switchgrass, indiangrass, and eastern gamagrass: current status and future. p. 159–170. *In* J.R. Carlson and E.D. McArthur (chair) Range plant improvement in western North America. Proc. of Symp., Annu. Meet., Salt Lake City, UT. 14 Feb. Soc. of Rang Manage., Salt Lake City.

Voight, P.W., and R.S. MacLauchlan. 1985. Native and other western grasses. p. 177–187. *In* M.E. Heath et al. (ed.) Forages. Iowa State Univ. Press, Ames.

Voight, P.W., and W.R. Oaks. 1985. Lovegrass, dropseeds, and other desert and subtropical grasses. p. 178–187. *In* J.R. Carlson and E.D. McArthur (chair) Range plant improvement in western North America. Proc. of Symp., Annu. Meet., Salt Lake City, UT. 14 Feb. Soc. of Range Manage. Salt Lake City.

4 Progress and Benefits to Humanity from Breeding Cool-Season Grasses for Turf[1]

William A. Meyer

Pure Seed Testing, Inc.
Hubbard, Oregon

C. Reed Funk

Rutgers University
New Brunswick, New Jersey

ABSTRACT

Dramatic progress continues to be made in the genetic improvement of a number of cool-season grasses used for turf and conservation. Attractive, durable, persistent cultivars with improved pest resistance, increased stress tolerance, and reduced maintenance requirements are becoming available to the American public. Such cultivars are becoming increasingly important in programs to enhance our physical environment, conserve our soil and water resources, and provide recreation. Advances in both the sciences and art of plant breeding and in associated disciplines are making turfgrass breeders more effective in meeting the needs and wants of humanity.

Turfgrass breeders continue to enjoy the excitement that comes from making new discoveries and developing clearly superior new cultivars. Landmark cultivars have been developed in many of our important turfgrass species. 'Merion' and 'Adelphi' Kentucky bluegrass (*Poa pratensis* L.), 'Manhattan' and 'Pennfine' perennial ryegrass (*Lolium perenne* L.), 'Rebel' tall fescue (*Festuca arundinacea* Schreb.), 'Biljart' hard fescue (*F. longifolia* Thuill.), 'Jamestown' Chewing's fescue (*F. rubra* L. ssp. *commutata* Gaudin), 'Bighorn' blue sheeps fescue (*F. ovina* L. ssp. *glauca*), 'Penncross' creeping bentgrass (*Agrostis palustris* Huds.), 'Tifgreen' bermudagrass (*Cynodon dactylon* Pers. × *transvaalensis* Burtt-Davy), and 'Meyer' zoysiagrass (*Zoysia*

[1] Joint contribution from Pure Seed Testing, Inc. and New Jersey Agric. Exp. Stn., Cook College, Rutgers Univ., publ. no. D15264-2-88. Some of this work was conducted as part of NJAES project no. 15264, supported by New Jersey Agric. Exp. Stn. funds, other grants, and gifts.

japonica Steud.) are examples of cultivars that have benefitted the turfgrass industry and stimulated additional breeding efforts.

Attributes of improved turfgrass cultivars include ease and economy of establishment, dependability, persistence, durability, reduced maintenance requirements and attractive appearance. Each of these attributes can be enhanced as we improve (i) resistance to diseases, insects, nematodes, and weeds; (ii) develop lower-growing grasses with better turf-forming properties; and (iii) increase tolerance of various environmental stresses such as heat, cold, and drought. Turfgrasses with improved tolerance of shade, close mowing, excessive wear, and poor soil conditions can be of great value. Cultivars of unique or striking color and appearance can be useful for special landscape patterns and effects.

KENTUCKY BLUEGRASS

Kentucky bluegrass is a hardy, persistent, attractive grass adapted to a wide range of soils and climates. Of the cool-season species, it has the most extensive rhizomes for sod strength and survival. Kentucky bluegrass provides nutritious and highly palatable grazing on more than 15 million ha of pastures throughout the northern half of the USA. As a premier turfgrass, it is used alone or in mixtures on more than 40 million lawns throughout all but the warmest regions of the USA and throughout Canada (Duell, 1985).

The value of Kentucky bluegrass for turf in humid regions was greatly enhanced with the discovery of 'Merion' by Joseph Valentine in the early 1930s. This highly apomictic cultivar originated as the progeny of a single plant growing on the Merion golf club in Ardmore, PA. After evaluation by agronomists at the USDA, the U.S. Golf Association Green Section, the Pennsylvania Agric. Exp. Stn., and progressive seed growers in the Pacific Northwest, Merion became available to the turfgrass grower in the early 1950s. Since then, more than 35 million kg of seed have been used in North America and Europe. This one cultivar revolutionized the cultivated sod industry and has had a tremendous impact on nearly all aspects of the turfgrass industry. In humid environments and under conditions of close mowing, turf of Common Kentucky bluegrass may show 60 to 100% turf loss from infection by the leaf spot and melting-out disease where adjacent plots of Merion are left virtually undamaged (Halisky et al., 1966).

Breeding Behavior

Kentucky bluegrass is a facultative apomict with complex cytological and embryological characteristics. The species has a highly variable chromosome number ($x = 7$; $2n = 28$–154) (Grazi et al., 1961; Akerberg, 1939; Muntzing, 1933; Tinney, 1940). It has the ability to hybridize with and absorb entire genomes from other species of *Poa* (Clausen, 1961). The ability of apomictic reproduction to facilitate fertility of superior hybrid genotypes helps explain the great genetic diversity and widespread distribution of Kentucky bluegrass. Different Kentucky bluegrass plants show a wide array of variation in the relative frequency of apomictic vs. sexual reproduction. This frequency of apomictic reproduction is primarily under genetic control with only modest variations occurring under normal environmental conditions (Hovin et al., 1976). Many Kentucky bluegrass cultivars originated as the progenies of single, highly apomictic plants selected from old turfs (Hintzen & van Wijk, 1985). Others resulted from the selection of naturalized ecotypes that consist of blends of plants showing varying degrees of apomixis. A few cultivars were developed by blending a number of highly apomictic single plant progenies. More recently, highly apomictic cultivars have been developed through controlled hybridization of parents showing varying degrees of apomictic reproduction (Bashaw & Funk, 1987).

Apomixis, as it exists in Kentucky bluegrass, is a nearly ideal way of maintaining superior hybrid genotypes. It combines the genetic advantages of asexual reproduction with the agronomic advantages of seed. As plant breeders understand and gain better control over apomictic reproduction, they will be able to develop superior hybrid cultivars more effectively.

Greenhouse hybridization techniques have been developed that permit the use of highly apomictic plants as female parents (Pepin & Funk, 1971; Hintzen & van Wijk, 1985). Programs using these techniques have developed a number of widely used cultivars including 'Adelphi', 'America', 'Classic', 'Midnight', and 'Nassau'.

Disease and Insect Resistance

Substantial success has been made in breeding for resistance to many important turfgrass diseases. Beginning with Merion's release in 1947, dozens of new Kentucky bluegrass cultivars have been developed with good resistance to the leafspot and melting-out disease incited by *Helminthosporium vagans* Drechsler. Genetic resistance to this disease has been quite stable. This may be partly due to the association of resistance with morphological characteristics such as wider leaves, a more decumbent growth profile, and the ability to maintain a leafy lower-growing turf under conditions of frequent close mowing. Resistance to the Helminthosporium leafspot and melting-out disease is critical for the production of a high-quality persistent turf in regions with cool, cloudy, and humid conditions.

Kentucky bluegrass cultivars have been developed with good resistance to many additional important diseases. These include (i) the stripe smut disease caused by *Ustilago striiformis* (Westend.) Niesl., (ii) leaf, stem, and stripe rusts caused by various species of *Puccinia*, (iii) powdery mildew caused by *Erysiphe graminis* D. C., and (iv) dollar spot incited by *Sclerotinia homoeocarpa* F. T. Bennett (now considered species of *Lanzia* and *Moellerodiscus* by some mycologists). However, increased stability of genetic resistance to many of these diseases is needed. Blending of resistant cultivars is helpful and occurs frequently in the turfgrass industry. Recent research on a number of patch diseases (Smiley & Craven-Fowler, 1984) should give breeders the knowledge and techniques needed for more efficient development of disease-resistant cultivars.

Kentucky bluegrasses have been identified with significant levels of resistance to damage from a number of important insect pests. These include a number of species of billbug (*Sphenophorus* spp.), the chinch bug (*Blissus leucopterus hirtus* Montandon), the greenbug aphid (*Schizaphis graminum* Rondani), and a number of lepidopterous species of sod webworm. Efforts are currently being made to identify and transfer useful Acremonium-type endophytes into elite cultivars of Kentucky bluegrass in hopes of enhancing insect resistance.

Shade Tolerance

Kentucky bluegrass cultivars have been developed with improved performance in shaded environments. Additional progress is needed for adequate performance in many situations. Tolerance of reduced light intensity and modified light quality are only part of the problem. Tree root competition, modified edaphic conditions, reduced air circulation, and a microenvironment favoring many diseases affect turf performance in shade (Beard, 1973). Where air circulation is good, the cooling effect of light shade can enhance bluegrass performance, especially in warmer regions. However, restricted air movement common in many shaded locations can result in increased heat and humidity. Shade tolerant grasses need adequate vigor to compensate for reductions in tiller, rhizome, and root production associated with reduced light. However, a reduced rate of vertical leaf elongation is also required to lessen photosynthate removed by mowing.

Reduced Mowing Requirement

Kentucky bluegrasses such as Midnight, 'Nugget', and America have been developed with a lower growth profile and a slower rate of vertical leaf elongation. A differential response to daylength, whereby cultivars exhibit short-day behavior during a longer portion of the growing season, appears to partly explain their lower growth habit. Turfgrasses developed for reduced mowing requirements must have an adequate establishment rate, be aggressive enough to crowd out weeds, and have the vigor to rapidly recover from injuries, diseases, insects, and environmental stresses. Grasses able to perform well at reduced soil fertility levels are also necessary if quality turf is to be achieved under reduced mowing frequencies.

High Seed Yield

Kentucky bluegrasses combining high seed yield with superior turf performance are rare. Early maturing erect, common types of Kentucky bluegrass have been identified that produce economical seed yields under dry land production in northern Idaho and eastern Washington. Unfortunatey, present cultivars of this type are highly susceptible to the Helminthosporium leaf spot and melting-out disease and perform poorly in many turfs. Improved common-type bluegrasses with moderate leaf spot resistance are being developed. They should be useful to the turfgrass industry. The first turf-type bluegrasses to be released, including Merion, had a low to moderate seed-yielding capability. Seed had to be produced with supplemental irrigation or in regions with higher rainfall because of their later reproductive maturity. The later introduction of 'Baron' and other turf-type bluegrasses with high seed yield and good turf quality has made seed yield an important criterion in turfgrass breeding programs. Ergot, caused by *Claviceps purpurea* (Fr.) Tul., is a serious disease in seed production fields of Kentucky bluegrass. No cultivars have been found to be highly resistant. There is a need for new cultivars resistant to this disease.

Tolerance of Heat, Drought, and Close Mowing

Kentucky bluegrasses with increased tolerance of heat and drought have been developed but substantial additional progress is needed. Many of the newer lower-growing turf-type bluegrasses are significantly more tolerant of the close mowing desired on golf course tees and fairways, especially in regions having cool summer nights and higher light intensities. However, the combined stresses of close mowing with the diseases and stresses found in warm humid climates make it difficult to maintain an attractive, close-cut persistent turf in such areas with presently available cultivars.

PERENNIAL RYEGRASSES

Perennial ryegrass is an important cool-season grass extensively used for both forage and turf in many areas of the world. The species is best adapted to areas with fertile soils and maritime climates having mild winters and cool moist summers. It has been cultivated for forage in Great Britain for at least three centuries where it became prominent in pastures, especially on rich, heavy soils of the lowlands. During this time numerous cultivars or strains were selected, ranging from short-lived stemmy types to the leafy persistent cultivars selected from old closely grazed grasslands. The short-lived, stemmy "hay-types," as they are referred to in Europe, are excellent seed producers. Large quantities of such cultivars including 'Linn', 'Nui', and 'Victoria' have been sold in inexpensive seed mixtures for turf use. Prior to the mid-1960s, this was the only type of perennial ryegrass generally available to the turfgrass industry of North America. Poor turf performance and limited persistence gave perennial ryegrass a bad name among turf growers prior to the development of improved turf-type cultivars.

Turf managers in the British Isles and much of Europe recognized the benefits of using the leafy, persistent "pasture types" for turf. These cultivars gave reasonably good performance for sports turf in cool maritime climates. They performed poorly throughout most of the USA due to lack of heat and cold tolerance and high susceptibility to diseases prevalent in warm, humid environments (Funk et al., 1985).

A turf-type ryegrass revolution began in the USA in the mid-1960s when Howard Kaerwer of Northrup King developed 'NK-100'. Robert Russell of J. and L. Adikes recognized the merits of NK-100 and promoted extensive use on Long Island, NY and nearby regions. Additional research by Howard Kaerwer showed that improved turf-type ryegrasses were extremely useful for the winter overseeding of dormant warm-season turfs in the southern USA. This started the overseeding revolution that has substantially improved southern lawns and sports fields for cool-season use.

A new era in ryegrass improvement for turf use was initiated when turfgrass personnel developed 'Manhattan' at the New Jersey Agric. Exp. Stn. in 1967 and Dr. Joseph Duich at Pennsylvania State Univ. developed 'Pennfine' in 1970. The ability of these ryegrasses to establish quickly and provide an attractive, leafy, persistent, wear-tolerant turf made them popular in many areas of the world. More than 18 million kg of Manhattan seed has been used in the USA and abroad. Manhattan has been listed as a recommended cultivar in the Netherlands, Sweden, Belgium, Denmark, France, and Germany, where it is still popular on soccer fields and other turfs receiving heavy wear. Pennfine, with its improved seed-yielding ability and better resistance to the Rhizoctonia brown patch disease also became an immediate success. The improved performance of Manhattan and Pennfine stimulated turfgrass breeders throughout the world to initiate or expand efforts to improve perennial ryegrass for turf. This has led to an abundance of excellent new ryegrasses. Leading cultivars currently include 'Palmer', 'Manhattan II', 'Gator', 'Fiesta II', 'Citation II', 'Omega II', 'All*Star', 'Cowboy', 'Premier', 'Pennant', 'Prelude', '246', 'Tara', 'Fiesta', 'Blazer', 'Derby', 'Repell', 'SR4000', 'SR4100', and 'Regal'. Promising new cultivars include 'Pinnacle', 'Commander', 'Dasher II', 'Pleasure', 'Sherwood', 'Aquarius', 'Regency', 'Allaire', 'Rodeo', 'Blazer II', 'Patriot', 'Saturn', 'Charger', and 'Dandy'. Improved turf-type perennial ryegrass use is increasing dramatically as superior cultivars become available. An estimated 28 million kg of turf-type perennial ryegrass seed is planted annually in the USA. At present, some of our most attractive and durable turf consists of pure stands of perennial ryegrass or mixtures of ryegrass with improved Kentucky bluegrasses.

Establishment

Rapid, easy establishment has been a prime factor in the popularity of improved turf-type ryegrasses with both homeowners and turfgrass professionals. With favorable conditions germination occurs within 5 d. A mowable turf can be produced within 3 wk with near-optimum temperature, moisture, and fertility. Rapid establishment reduces weed problems and enhances erosion control. Successful seedings can be made during most months of the growing season if properly managed.

Growth Characteristics, Texture, and Density

The improved turf-type perennial ryegrasses have the ability to produce an attractive, leafy, medium dense turf in their area of adaptation. Present cultivars are substantially lower-growing than the common perennial ryegrass types such as Linn. New lower-growing "dwarf-types" are presently being developed. Such cultivars provide an exceptionally attractive, dense turf with a reduced mowing requirement especially under cool short daylength environments. They should be especially useful for the winter overseeding market. With additional improvements in summer stress tolerance, their area of usefulness will be expanded.

Thatch

Turf-type ryegrasses produce little or no thatch. This lack of excessive thatch buildup enhances both long-term performance and the ease of reseeding damaged turf.

Color

The yellow-green color of common perennial ryegrass is usually too light to blend well with Kentucky bluegrass and fine fescues. Most turf-type ryegrasses developed in the USA have an attractive, bright, medium dark green color. However, cultivars ranging from the medium green color of 'Loretta' and 'Elka' to the dark bluegreen color of '246' and 'Caravelle' are available.

Wear Tolerance

Manhattan and other improved turf-type perennial ryegrasses are used on soccer fields in northern Europe because of their excellent wear tolerance, rapid recovery from injury, ability to compete with *Poa annua*, and good winter performance. Wear tolerance is greatest during their optimum periods of growth during the cool, moist periods of spring and fall. They have a low crown, a tough leaf, and an excellent capacity for recovery. Excellent establishment characteristics allow for successful overseeding during the playing season on areas subjected to excessive use.

Mowing Characteristics

Ryegrasses traditionally develop tough leaves that may be difficult to mow neatly under stress conditions and during the reproductive period of late spring. The improved turf-type perennial ryegrasses show substantial improvements in mowing quality. Most of the newer cultivars can be mowed cleanly and with little difficulty when growing conditions are favorable. The more heat-tolerant cultivars show less loss of mowing quality during the summer months. Latter maturing ryegrasses are less likely to develop excessive stemminess during their reproductive period.

Shade Tolerance

Turf-type perennial ryegrasses perform better in shade than common perennial ryegrasses. Ryegrasses growing in cool shade have softer leaves that mow with greater ease. Although present cultivars will not persist in excessively heavy shade, they can provide good temporary turf. Their rapid establishment allows them to produce several months of turf cover during periods when tree leaves are absent or just developing. They will persist several weeks into the summer after the leaves have returned. Breeding programs for additional improvements in shade tolerance should be encouraged.

Summer Performance

The best new turf-type perennial ryegrasses demonstrate substantial improvements in heat tolerance and resistance to summer diseases. Increased resistance to the brown patch disease caused by *Rhizoctonia solani* Kuhn is an extremely important factor in improved performance in areas with warm, humid summers. The earlier ryegrass cultivars developed in the cool summer climates of the British Isles and northwestern Europe are highly susceptible to Rhizoctonia brown patch. Newer turf-type cultivars originating from germplasm collected from old turfs of the mid-Atlantic region of the USA show significantly improved resistance to this disease. Cultivars with higher levels of resistance are being developed using population improvement techniques. It is also of interest to develop ryegrasses with reduced heat tolerance for some winter overseeding programs in the south.

Winter Performance

Most turf-type perennial ryegrasses are substantially more winter hardy than common perennial ryegrass. Vermont tests show this to be associated with a low crown and a reduced subcrown internode (Cohen, 1980). Improved resistance to many winter diseases, including snow molds and winter net blotch, has been achieved in ryegrass breeding programs.

Seed Production

Excellent yields of high-quality seed are being achieved with many of the best new turf-type perennial ryegrasses. This is saving consumers millions of dollars each year. Genetic resistance to stem rust (incited by *Puccinia graminis* Pers.) has been incorporated into many improved ryegrasses (Rose-Fricker et al., 1986). This has substantially reduced the need for fungicides normally used in ryegrass seed production. Multiple genes for rust resistance discovered in ryegrasses showing poor turf qualities were backcrossed into the best turf-type germplasm pools concurrent with other population improvement programs to increase resistance to other diseases, and improve stress tolerance, seed yield, and turf performance.

Endophyte-Enhanced Performance

Acremonium endophytes have been incorporated into many new cultivars of turf-type perennial ryegrasses. Although this endophytic fungus is associated with occasional outbreaks of ryegrass staggers in sheep (*Ovis aries*) and some other grazing animals, it is generally considered a desirable attribute in grasses grown for turf and conservation. New Zealand scientists (Prestidge et al., 1982; Mortimer et al., 1982) were the first to report an association between the endophytic fungus, *Acremonium lolii*, and resistance to the Argentine stem weevil, *Listronotus bonariensis*, in field trials of perennial ryegrass. Subsequent studies (Funk et al., 1983; Ahmad et al., 1986) reported that many instances of resistance to species of lepidopterous sod webworm and billbugs (*Sphenophorus* spp.) were associated with the presence of the Acremonium endophyte. We have observed instances of better summer survival, enhanced fall recovery, and reduced weed invasion in perennial ryegrass, tall fescue, hard fescue, and Chewings fescue turfs containing high percentages of endophyte-infected plants (Funk et al., 1985). To date, we have not seen any adverse effects of endophytes on turfgrass performance. Endophytes appear to have little or no effect on turf when growing conditions are favorable. Observations of endophyte-enhanced performance have usually occurred under conditions of severe heat or drought stress or attacks by some, but not all, insect pests. Endophytes in turfgrasses may be similar to insurance; of little value when conditions are favorable, but of considerable value under certain conditions of environmental and biological stress. An increased knowledge of the role of endophytes is making cultivar evaluation and breeding programs much more effective.

FINE FESCUES

Fine fescues include a number of fine-leafed species of *Festuca* (Schmit et al., 1974). They have fine, tough, bristle-like leaves, and a leafy turf-type growth habit. They are tolerant of acid, infertile, droughty soils and moderate shade but intolerant of poor drainage and high N especially during hot weather.

Chewings fescue (*F. rubra* L. ssp. *commutata* Gaud.) is a fine-leafed, low-growing grass lacking rhizomes. In cool summer climates, improved cultivars can produce an attractive turf requiring limited fertilizer and are tolerant of rather close mowing. Turfgrass breeders in Europe and North America have produced a number of attractive cultivars with improved resistance to many important diseases and good seed-yielding ability. Current programs are directed to increasing heat tolerance and improving disease and insect resistance. Acremonium-type endophytes have been identified in Chewings fescue that enhance resistance to a number of insect pests (Saha et al., 1987). Useful endophytes are currently being incorporated into a number of leading cultivars and elite germplasms using the backcross technique.

Strong creeping red fescue (*F. rubra* L. ssp. *rubra*) have abundant, long, spreading rhizomes; have larger seed and better seedling vigor; and generally produce less thatch than the Chewings fescues. They are also more compatible in mixtures with the Kentucky bluegrasses and turf-type perennial ryegrasses. Substantial progress has been made in the development of cultivars with high seed yield potential, a lower turf-type growth profile, and good resistance to powdery mildew incited by *Erysiphe graminis*. Only moderate success has been achieved in obtaining cultivars with good resistance to net-blotch (caused by *Drechslera dictyoides* f. sp. *dictyoides* Drechs), red thread [caused by *Laetisaria fuciformis* (McAlp.) Burdsall] and dollar spot (incited by *Sclerotinia homoeocarpa* Bennett). Efforts are being made to find useful nonchoke-inducing Acremonium-type endophytes in the strong creeping red fescues.

Improved cultivars of slender creeping red fescues [*F. rubra* L. ssp. *litoralis* (Meyer) Auquier] can produce a fine dense, compact, turf with short, slender rhizomes. Specialized cultivars have been developed with excellent tolerance of saline soils ('Dawson') and specific heavy metals ('Merlin' and 'Sterling') (Shildrick, 1980; McHugh et al., 1985). Programs are underway to improve seed yield and resistance to red thread and dollar spot.

Improved cultivars of hard fescue (*F. longifolia* Thuill.) have been developed that show exceptional promise for low maintenance turfs. Improved hard fescues produce a turf comparable in texture and density to the better Chewings fescue cultivars but with a reduced rate of vertical growth, lower fertility requirements, and better resistance to red thread, net blotch, and dollar spot (Meyer, 1982). Turfgrass breeders are currently attempting to improve seed production efficiency, seedling vigor, and speed of recovery from injury. Useful nonchoke-inducing Acremonium-type endophytes are being introduced into many leading cultivars of hard fescue to enhance insect resistance and perhaps stress tolerance.

Blue sheeps fescue (*F. ovina* L. ssp. *glauca*) is a durable turfgrass often found in shade and on poor soils receiving low maintenance. The cv. Bighorn demonstrates that attractive, dense, low-growing varieties can be developed with good shade tolerance, good disease resistance and low maintenance requirements. Nonchoke-inducing Acremonium-type endophytes have been discovered in blue sheeps fescue. Studies have been initiated to determine their usefulness in enhancing insect resistance and improving turf performance.

TURF-TYPE TALL FESCUE

The availability and usage of improved tall fescues for turf has increased dramatically in the past 5 to 7 yr in the USA. The present U.S. seed production is estimated at 20 million kg yearly. This species has been an important grass for forage, roadside stabilization, and erosion control for many years. Compared to other popular cool-season species, tall fescue possesses a much deeper root system that provides drought tolerance or avoidance. This species also has the ability to persist at reduced soil fertility and has improved tolerance to many insects compared to other cool-season species (Buckner & Bush, 1979).

Tall fescue is a native of Europe and adjacent regions and was introduced into the USA during the 1800s. It did not become widely used in the USA until the development and release of 'Alta' and 'Kentucky 31' in 1940 and

1943, respectively (Hanson, 1965). These two cultivars were developed primarly as forage grasses. By the 1960s, states in the transition zone of the USA were recommending Kentucky 31 tall fescue for turf because of its wide adaptation to variable soil pH, rainfall, and sunlight. Kentucky 31 and Alta have both been used widely for turf in southwestern USA and at higher elevations of Arizona and Nevada.

Tall fescue is basically a bunchgrass, although short rhizomes may occasionally occur on isolated plants in sandy soils. Named cultivars of tall fescue in the USA are cross pollinated and possess $2n = 6x = 42$ chromosomes. It has a good seed germination rate with emergence in 6 to 7 d in warm soils. Tall fescue does not tiller as rapidly as perennial ryegrass so higher seeding rates are required.

New Improved Cultivars

Since 1972, breeding programs were initiated to develop cultivars of tall fescue more suited for turf use. Most of the new cultivars have been derived from germplasm selected from old turf areas in the northeastern and southeastern USA. In naturalized stands throughout the country, a small percentage of plants have been found that have a lower growth habit, finer leaves, and a reduced rate of vertical growth. 'Rebel' was the first turf-type cultivar (Funk et al., 1981). It was developed by the New Jersey Agric. Exp. Stn. 'Falcon' and 'Olympic' followed closely behind as new improved cultivars.

These cultivars have leaves approximately 30 to 40% finer than Kentucky 31 and tiller densities almost twice that of Kentucky 31. They also have a darker green color, improved persistence under close mowing and better overall disease resistance compared to Kentucky 31. Other improved turf-type cultivars include 'Bonanza', 'Adventure', 'Jaguar', 'Apache', 'Arid', 'Mustang', 'Rebel II', 'Houndog', 'Finelawn 5GL', 'Baja', 'Trailblazer', 'Cimmaron', 'Monarch', and 'Mesa'. Seed production is being initiated on a number of newer cultivars including, 'Wrangler', 'Tribute', 'Silverado', 'Eldorado', 'Jaguar II', 'Murietta', and 'Chieftain'.

Turf-type tall fescues have been widely used to improve the appearance and durability of homelawns, playgrounds, athletic fields, and parks throughout many areas of the USA. The reduced irrigation and fertility requirements have made it possible to greatly improve turf quality with reduced energy inputs.

Lower-growing Dwarf-type Cultivars

Lower-growing turfgrasses have a great appeal in efforts to decrease mowing costs and improve performance. The first turf-type tall fescue cultivars had a deeper crown and a significantly lower leaf canopy. However, the normal mature plant height of these cultivars, including Falcon and Rebel, range from 130 to 145 cm. In breeding populations of the new turf-type tall fescues, a small percentage of plants have been found that are 100 cm or less. These lower-growing dwarf-types also show a reduced rate of vertical growth and higher tiller densities than Rebel or Falcon when grown in turf.

The first dwarf-type cultivar released was 'Trailblazer'. There are many new promising experimental tall fescues that exhibit a similar compact dense growth habit. These selections have a reduced tendency to scalp when mowing is irregular because of weather conditions.

Because these new dwarf-types are dense, there is a concern that under humid, hot conditions they will be under greater disease pressure. Also, the high density could result in reduced drought tolerance. These new dwarf-types have performed exceptionally well under irrigation in arid climates such as California when compared to other cool-season species.

Lower-growing tall fescues also have the advantage of reduced lodging in seed production fields. The new dwarf-type cultivars have resulted from four or five cycles of phenotypic recurrent selection for a shorter mature height, increased tiller number, and improved disease resistance.

Disease Resistance

The most common and serious diseases of tall fescue are brown patch, caused by *Rhizoctonia solani* Kuhn, and net blotch, caused by *Drechslera dictyoides* f. sp. *dictyoides* Drechs (Clarke et al., 1986; Meyer, 1982). Through recurrent selection and single plant progeny turf evaluations, progress has been made in developing cultivars of turf-type tall fescue with improved resistance to the above two diseases. Silverado, Jaguar II, Bonanza, and Monarch have shown the best resistance to net blotch to date.

Even though many cultivars have shown improvements in brown patch resistance more development work is needed. This disease is especially severe when moderate drought stress proceeds hot and humid conditions favoring disease development. Cultivars with improved drought stress resistance frequently show less damage from this disease.

Crown rust, caused by *Puccinia coronata* Cda., can be severe on tall fescues in some areas of the USA, especially southern California. The cv. Fawn and Alta are both susceptible to most races of crown rust (Meyer, 1982), while Kentucky 31 and most of the new turf-type cultivars have shown improvements in resistance to most races of this fungus. Cultivars with a broad base of improved resistance are still needed.

Stem rust, caused by *Puccinia graminis*, has been a serious disease of perennial ryegrass in seed production fields in western Oregon for the past 20 yr. Although stem rust has been observed on tall fescue in a few locations throughout the USA in recent years, it had not become a serious problem in seed production fields in western Oregon. However, in the spring of 1987, a few seed fields were seriously damaged with severe reduction in seed yield. Preliminary observations in old nurseries of tall fescue indicate that some of the present turf-type cultivars contain plants with resistance to this disease. Breeding for resistance to this potentially damaging disease is of high priority.

A number of *Pythium* spp. can cause devastation to seedlings and mature turf of tall fescue under hot, humid, wet conditions. No resistance to this disease has been reported among the presently available cultivars. Sources of resistance should be found to this disease.

Stress Tolerance

The tall fescues are adapted to a wide range of soil types. They have above-average tolerance of acidic, alkaline, saline, and poorly drained soils. Tall fescues have a deep root system with better drought and heat tolerance than ryegrasses, bluegrasses, fine fescues, and bentgrasses. The root systems of the turf-type cultivars available prior to 1985 appear comparable to those of Alta and Kentucky 31 (Kopec, 1985).

The new turf cultivars have shown good shade tolerance compared to Kentucky 31. These new cultivars have shown finer leaf blades in shade and have performed as well or better than the other cool-season species. Tall fescue is not as winter-hardy as Kentucky bluegrass in some northern areas of the USA, especially under high N fertility and ice sheeting conditions. The new turf-type cultivars have shown good traffic tolerance compared to Kentucky 31 and Alta. Once they are well established, they have shown good recuperative abilities and the ability to grow in moderately compacted soils.

Insect Resistance

The tall fescue species has good overall tolerance to insects compared to other cool-season grass species. In many trial locations, the tall fescues have not been damaged by billbugs and sod webworms while ryegrasses, bluegrasses, and fine fescues have suffered severe damage in adjoining plots.

Seed Yield

The new turf-type tall fescues have been comparable or better seed yielders than earlier cultivars. Recurrent selection programs in western Oregon have resulted in cultivars with uniform maturity, a high proportion of reproductive tillers and improved resistance to lodging.

One of the problems in tall fescue seed production is the depression of seed yields in older (3–5 yr) fields, where reproductive development is suppressed by sod bounding and intertiller competition. Field burning can be used to thin older fields, but this activity is being curtailed because of environmental concerns. There is a need for cultivars with improved ability to maintain high seed yields in older stands.

Crown rust can cause a slight reduction in tall fescue seed yields. In 1987, stem rust was seen to have a severe effect on seed production. A temporary solution will be the use of fungicides to control stem rust until cultivars with good resistance can be developed. Observations in western Oregon since 1977 have shown that early maturing cultivars have better seed-yielding ability than later-maturing cultivars.

Current Challenges and Future Prospects

Future breeding in tall fescue for turf should be carried out in at least two separate directions. There is a need for very fine, dwarf, dense cultivars to be grown in arid Mediterranean-type climates with irrigation as part of the management. These types will have reduced mowing requirements and the potential for high-quality turf.

Another breeding direction should be for highly stress-resistant cultivars in terms of drought and heat tolerance along with high levels of brown patch resistance. Selection for high levels of insect and Pythium resistance is also needed.

In the seed-producing areas, there will be a need for cultivars with improved stem rust resistance to improve the economics of seed production. Some of these breeding programs should be in the absence of field burning which is becoming more restricted each year.

BREEDING BENTGRASSES

There are three species of bentgrass that have had the widest usage for closely mowed turf in the USA. Creeping bentgrass (*Agrostic palustris* Huds) is the most important species, colonial bentgrass (*A. tenuis* Sibth.) is the second most important species, and velvet bentgrass (*Agrostis canina* L.) is a minor species.

CREEPING BENTGRASS

Creeping bentgrass, native to Eurasia, is a unique cool-season species that exhibits a vigorous stoloniferous growth habit. It forms its best quality turf at cutting heights of 1.5 cm or less. Creeping bentgrass is the primary cool-season grass used for golf course putting greens, tees, and closely mowed fairways in the USA. It can also be a difficult-to-control weedy grass if it invades other cool-season species such as Kentucky bluegrass and perennial ryegrasses mowed at 3 to 6 cm (Beard, 1973).

Many of the early cultivars of creeping bentgrasses for fine putting greens were vegetatively propagated. 'Congressional', 'Toronto', 'Washington', and 'Cohansey' were four of the most popular cultivars. These were all clonal selections from old turf areas (Beard, 1973).

The two most popular cultivars of seeded creeping bentgrass are 'Seaside', released in the 1920s (Hansen, 1965) and 'Penncross', released in 1954 (Hein, 1958). Seaside is a collection from Oregon and is a non-uniform cultivar prone to segregation into mottled patches. Many of the plants in Seaside are coarse in leaf texture.

Penncross is a first generation synthetic cultivar produced by the random crossing of three vegetatively propagated strains selected by H.B. Musser of Pennsylvania Agric. Exp. Stn. (Hein, 1958). This landmark cultivar is still the most widely used grass for putting greens and tees in the cool-season growing areas of the USA. Penncross has superior vigor, heat and traffic tolerance when compared to other bentgrasses. It is an aggressive grower with good recuperative potential, but a proneness to thatch if managed improperly.

The cv. 'Prominent' and 'Emerald' were developed in Europe and have only had limited usage. 'Penneagle' was released in 1978 as a less-aggressive grass than Penncross (Duich, 1979). It has been widely used for closely cut fairways in the USA and has shown better resistsance to dollar spot in some trials. Penneagle appears to be somewhat less tolerant to heat and traffic than Penncross at the extremely close cutting heights used on many putting greens.

'PennLinks' is another new cultivar released in 1987 by Dr. J. M. Duich of Pennsylvania State Univ. This cultivar produces a high-quality fine, dense putting green turf with the least amount of genetic segregation when compared to other available cultivars. Heat tolerance appears to be improved while reactions to disease appear similar to other improved cultivars.

All of the presently available cultivars of creeping bentgrass are susceptible to an array of fungal disease problems. Fungicide and insecticide applications are commonly used on fine turf areas. There is a great challenge facing the bentgrass breeder to improve the disease resistance of this species.

Every one of the presently available bentgrass cultivars are late in maturity compared to other cool-season species. This results in high seed production costs because of extensive irrigation requirements. There is a need to find bentgrass germplasm with a much earlier maturity to be able to produce seed more efficiently.

COLONIAL BENTGRASS

Colonial bentgrass is a native of Europe and consists of plants with varied amounts of stoloniferous and rhizomatous growth. This species does have less stolon development than creeping bentgrass but usually crowds out other cool-season grasses such as Kentucky bluegrasses or perennial ryegrasses in polystands.

'Astoria' and 'Highland' bentgrass are cultivars developed prior to 1930. They were ecotype collections from naturalized stands in western Oregon (Hanson, 1965). Highland has been the most widely available cultivar. This cultivar is susceptible to many fungal diseases, has poor drought tolerance and gets puffy when mowed at normal to higher cutting heights. Other cultivars are 'Exeter' and 'Holfior' (Hanson, 1965). These cultivars have performed somewhat better than Highland, but need improvements in disease resistance and stress tolerance.

A cultivar is needed with no stolon growth and good rhizome development. These types would be more compatible in mixtures with other cool-season species and would have better recuperative potential from traffic and divots.

VELVET BENTGRASS

Velvet bentgrass (*Agrostis canina* L.) is a minor species of bentgrass compared to colonial or creeping bentgrass. This species produces an extremely fine-textured turf with a velvety appearance (Beard, 1973). The stolon development is intermediate to creeping and colonial bentgrass. This species has been used in cooler areas of the cool, humid regions of the USA, primarily New England. It is prone to wilt and is susceptible to many important disease problems. Velvet bentgrass performs well in a shaded area. 'Kingston' is available in limited quantities as an improved seeded cultivar.

ROUGH BLUEGRASS

Improved turf-type cultivars of rough bluegrass (*Poa trivialis* L.) can be useful for certain specialty turfs. They are best adapted for permanent turf use in cool, moist, shaded environments in temperate climates and in mixtures for the winter overseeding of dormant warm-season grasses in the south (Hurley & Funk, 1985). Rough bluegrass is a shallow-rooted, stoloniferous species with excellent shade tolerance in cool moist locations. It has excellent cold tolerance and the ability to germinate and grow in cool weather. However, it is intolerant of heat and drought stress. These characteristics make rough bluegrass incompatible in turfgrass mixtures grown in full sun in temperate regions. Improved turf-type cultivars such as 'Colt' and 'Sabre' have been developed with a darker green color, greater density, finer texture and the ability to produce a lower growing, more compact turf (Dickson et al., 1980). Improved resistance to shattering would make seed production more reliable and economical.

FUTURE PROSPECTS

The importance of cool-season turfgrasses in soil preservation and improvement, their value in enhancing our environment, and their economic role in a multibillion dollar turfgrass industry dictates continued and expanded efforts in their genetic improvement. Turfgrass breeding is still in its infancy or early childhood. Much remains to be accomplished in the collection, evaluation, and preservation of germplasm. Research is needed to develop more efficient breeding and evaluation techniques. This will require basic studies in the nature and control of apomixis, an understanding of various host-pathogen interrelationships, and the responses of the grass plant to various environmental stresses. The roles of endophytic fungi and vascular-arbuscular endomycorrhiza in enhancing host plant performance needs additional study. Continued efforts of scientists in many disciplines should lead to a continuing stream of vastly improved cultivars for the benefit of all people.

REFERENCES

Ahmad, S., J.M. Johnson-Cicalese, W.K. Dickson, and C.R. Funk. 1986. Endophyte-enhanced resistance in perennial ryegrass to the bluegrass billbug, *Sphenophorus parvulus*. Entomol. Exp. Appl. 41:3–10.

Akerberg, E. 1939. Apomictic and sexual seed formation in *Poa pratensis* L. Hereditas 25:359–370.

Bashaw, E.C., and C.R. Funk. 1987. Apomictic grasses. p. 40–82. *In* W.R. Fehr (ed.) Principles of cultivar development. Vol. 2. Crop species. Macmillan Publ. Co., New York.

Beard, J.B. 1973. Turfgrass: Science and culture. Prentice Hall, Englewood Cliffs, NJ.

Buckner, R.C., and L.P. Bush (ed.). 1979. Tall fescue. Agronomy 20.

Clarke, B.B., C.R. Funk, and P.M. Halisky. 1985. Development of *Festuca arundinacea* Schreb. cultivars with improved resistance to *Rhizoctonia solani* Kuhn. p. 641–646. *In* F. Lemaire (ed.) Proc. 5th Int. Turf Res. Conf., Avignon, France. 1–5 July. INRA Publ., Versailles, France.

Clausen, J. 1961. Introgression facilitated by apomixis in polyploid *Poas*. Euphytica 10:87–94.

Cohen, R.P. 1980. Evaluating perennial ryegrass, *Lolium perenne* L., for cold hardiness through alcohol bath freezing and measurement of subcrown internode. Masters thesis. Univ. of Vermont, Burlington.

Dickson, W.K., G.W. Pepin, R.E. Engel, and C.R. Funk. 1980. Registration of Sabre roughstalk bluegrass. Crop Sci. 20:668.

Duell, R.W. 1985. The bluegrasses. p. 188–197. *In* M.E. Heath et al. (ed.) Forages: The science of grassland agriculture. 4th ed. Iowa State Univ. Press, Ames.

Duich, J.M. 1979. Penneagle creeping bentgrass. USGA Green Section Record. July/August, p. 8–10. U.S. Golf Assoc.,

Funk, C.R., W.K. Dickson, E. Clark, and J.M. Johnson-Cicalese. 1985. Performance and use of ryegrasses for forage and turf. Rutgers Turfgrass Proc. 16:86–100.

Funk, C.R., R.E. Engel, W.K. Dickson, and R.H. Hurley. 1981. Registration of Rebel tall fescue. Crop Sci. 21:632.

Funk, C.R., P.M. Halisky, S. Ahmad, and R.H. Hurley. 1985. How endophytes modify turfgrass performance and response to insect pests in turfgrass breeding and evaluation trials. p. 127–145. *In* F. Lemaire (ed.) Proc. 5th Int. Turf Res. Conf., Avignon, France. 1–5 July. INRA Publ., Versailles, France.

Funk, C.R., P.M. Halisky, M.C. Johnson, M.R. Siegel, A.V. Stewart, S. Ahmad, R.H. Hurley, and I.C. Harvey. 1983. An endophytic fungus and resistance to sod webworms: Association in *Lolium perenne* L. Bio/Technology 1:189–191.

Grazi, F., M. Umaerus, and E. Akerberg. 1961. Observations on the mode of reproduction and the embryology of *Poa pratensis* L. Hereditas 47:489–541.

Halisky, P.M., C.R. Funk, and R.E. Engel. 1966. Melting-out of Kentucky bluegrass varieties by *Helminthosporium vagans* as influenced by turf management practices. Plant Dis. Rep. 50:703–706.

Hanson, A.A. 1965. Grass varieties in the United States. USDA Agric. Handb. 170. U.S. Gov. Print. Office, Washington, DC.

Hein, M.A. 1958. Registration of Penncross creeping bentgrass. Agron. J. 50:399.

Hintzen, J.J., and A.J.P. van Wijk. 1985. Ecotype breeding and hybridization in Kentucky bluegrass. p. 213–219. *In* F. Lemaire (ed.) Proc. 5th Int. Turf Res. Conf., Avignon, France. 1–5 July. INRA Publ., Versailles, France.

Hovin, A.W., C.C. Berg, E.C. Bashaw, R.C. Buckner, D.R. Dewey, G.M. Dunn, C.S. Hoveland, C.M. Rincker, and G.M. Wood. 1976. Effects of geographic origin and seed production environments on apomixis in Kentucky bluegrass. Crop Sci. 16:635–638.

Hurley, R.H., and C.R. Funk. 1985. Genetic variability in disease reaction, turf quality, leaf color, leaf texture, plant density and seed shattering of selected genotypes of *Poa trivialis*. p. 221–226. *In* F. Lemaire (ed.) Proc. 5th Int. Turf Res. Conf., Avignon, France. 1–5 July. INRA Publ., Versailles, France.

Kopec, D.M. 1985. Tall fescue soil moisture depletion, evapotranspiration and growth parameters. Ph.D. diss. Univ. of Nebraska, Lincoln.

McHugh, S., M.S. Johnson, T. McNeilly, M.O. Humphreys, and R. Rowling. 1985. Tolerance to salt in *Festuca rubra* L. p. 185–193. *In* F. Lemaire (ed.) Proc. 5th Int. Turf Res. Conf., Avignon, France. 1–5 July.

Meyer, W.A. 1982. Breeding disease resistant cool-season turfgrass cultivars for the United States. Plant Dis. 66:341–344.

Mortimer, P.H., L.R. Fletcher, M.E. di Menna, I.C. Harvey, G.S. Smith, G.M. Barker, R.T. Gallasher, E.P. White. 1982. Recent advances in ryegrass staggers. p. 71–74. *In* Proc. of the 34th Ruakura Farmers' Conf., New Zealand.

Muntzing, A. 1933. Apomictic and sexual seed production in *Poa*. Hereditas 17:131–154.

Pepin, G.W., and C.R. Funk. 1971. Intraspecific hybridization as a method of breeding Kentucky bluegrass for turf. Crop Sci. 11:445–448.

Prestidge, R.A., R.P. Pottinger, and R.M. Barker. 1982. An association of Lolium endophyte with ryegrass resistance to Argentine stem weevil. p. 119–122. *In* Proc. of the 35th NZ Weed and Pest Control Society Conf.

Rose-Fricker, C.A., W.A. Meyer, and W.E. Kronstad. 1986. Inheritance of resistance to stem rust (*Puccinia graminis* subsp. *graminicola*) in six perennial ryegrass (*Lolium perenne* L.) crosses. Plant Dis. 70:678–681.

Saha, D.C., J.M. Johnnson-Cicalese, P.M. Halisky, M.I. van Heemstra, and C.R. Funk. 1987. Occurrence and significance of endophytic fungi in the fine fescues. Plant Dis. 71:1021–1024.

Schmit, R.M., R.W. Duell, and C.R. Funk. 1974. Isolation barriers and self-compatibility in selected fine fescues. p. 9–17. *In* E.C. Roberts (ed.) proc. 2nd Int. Turfgrass Res. Conf., ASA and CSSA, Madison, WI.

Shildrick, J.P. 1980. Species and cultivar selection. p. 69–97. *In* I.R. Rorison and R. Hunt (ed.) Amenity grassland—An ecological perspective. John Wiley and Sons, New York.

Smiley, R.W., and M. Craven-Fowler. 1984. *Leptosphaeria korrae* and *Phialophora graminicola* associated with Fusarium blight syndrome of *Poa pratensis* in New York. Plant Dis. 68:440–442.

Tinney, F.W. 1940. Cytology of parthenogenesis in *Poa pratensis*. J. Agric. Rev. 60:351–360.

5 Progress and Benefits to Humanity from Breeding Warm-Season Grasses for Turf

Philip Busey

University of Florida
Fort Lauderdale, Florida

ABSTRACT

Warm-season grasses that are used for turf are primarily in the genera *Axonopus, Cynodon, Eremochloa, Paspalum, Stenotaphrum,* and *Zoysia.* People in subtropical areas benefit from the recreation, aesthetic, sanitation, and conservation features that those grasses provide. The most widely used warm-season turfgrasses are cultivars selected by public plant breeders, although private efforts have been notable in development and marketing. Breeding has created warm-season turfgrass cultivars that are more utilitarian and less costly to maintain, compared with unselected common strains. Successes of breeding include selection for finer texture (e.g., 'Tifgreen' bermudagrasses, *Cynodon* × *magenissii* Hurcombe); for insect resistance [e.g., 'Floratam' St. Augustinegrass, *Stenotaphrum secundatum* (Walt.) Kuntze]; and for higher apparent density (e.g., 'Argentine' bahiagrass, *Paspalum notatum* Flügge). Most progress in breeding warm-season turfgrasses has been based on serendipitous creation, and skillful discovery. Further stepwise improvement from past successes is difficult, because there is generally no source germplasm available. Scientists have broadened the knowledge of warm-season grass genetics, but knowledge is still fragmentary in other critical areas. The accurate description and identification of cultivars and their wild relatives are almost nonexistent. Shade tolerance and acceptability under reduced irrigation are of the utmost importance, but have been vexing research problems. Pest resistance, while desirable, must be understood in the context of generalized (horizontal) adaptation to the stress environment. Not enough work has been done to correlate laboratory observed resistance with field level resistance. Future progress will depend upon better knowledge of taxonomy and genetic resources, selection method, and the basis of cultivar adaptation.

The improvement of warm-season turfgrasses began recently, but gives far-reaching benefits. People in subtropical regions have outdoor carpets, where once they had weeds and bare soil. Cultivars such as 'Tifway' bermudagrass (*Cynodon* × *magenissii* Hurcombe) provide for safe recreation. Insect-

 Contributions from Breeding Forage and Turf Grasses, CSSA Special Publication no. 15.

resistant cultivars such as 'Floratam' St. Augustinegrass [*Stenotaphrum secundatum* (Walt.) Kuntze] counteract the need for pesticides. Conservation cultivars such as 'Argentine' bahiagrass (*Paspalum notatum* Flügge) protect public easements from erosion and vegetation succession. These are a few of the successes of warm-season turfgrass breeding, which occurred basically since the late 1940s. As I will show, much more must be done.

There were minimal efforts to grow warm-season grasses, least of all breed them for turf, before 1900. Flint (1903) described the situation in 1887,

> Agriculturists at [sic] the South have scarcely been successful in the attention hitherto given to the introduction of valuable grasses . . . The normal range of the grasses, strictly speaking, is not so far south . . . It has been found extremely difficult to form a close turf or sward below the latitudes of the more equable distribution of rain . . . south of Baltimore . . . Most of the higher English grasses fail, in such situations, to form a close turf, and give place to the tufted or jungle grasses . . .

Flint's comments lead us to look at the social and biological environment of warm-season turfgrass breeding.

The transfer of European technology to the subtropics and warm temperate regions required modifications. Warm-season turfgrasses had to be modified genetically to serve the needs of newcomers. That is what this chapter is about. The culture of turf was a custom that first had to be introduced, because it had been developed in northern Europe and temperate east Asia. Finally, there had to be (and must continue to be) acceptance by newcomers of the potential benefits of warm-season turfgrasses. Immigrants to subtropical areas are still surprised by the coarseness of commonly grown warm-season grasses. Major textbooks emphasize that turfgrass quality is based on deep color and fine texture (narrow leaf blades). These are characteristics for which most warm-season grasses compare unfavorably, and this fact may have contributed to a delayed recognition of the value of warm-season grasses, and of their subsequent improvement for turf.

Warm-season turfgrasses differ from cool-season turfgrasses in having a four-carbon (C_4) molecule, contrasting with three-carbon, as the first step in carbohydrate assimilation. Their C_4 physiology is associated with low CO_2 compensation concentration (Krans et al., 1979). This ancient adaptation provides high growth rate at high light levels and at high temperatures. Warm-season turfgrasses have 69% the water consumption of cool-season turfgrasses, when grown under the same conditions (Biran et al., 1981). Warm-season grasses also have relatively poor adaptation to shade, a problem for growing turf in urban areas.

More heavily developed vascular bundles containing an inner bundle sheath layer and an outer mesophyll layer (Kranz leaf anatomy) may provide C_4 grasses a deterrence to some kinds of herbivores. The lower protein content of C_4 grasses is also correlated with a lower Homoptera load in a semiarid grassland in Texas (Boutton et al., 1978). Despite these physiological differences, warm-season turfgrasses have major insect pest problems, in part because of the long growing season in subtropical areas. Warm-season grasses are generally tougher to cut with a mower, and generally have a lighter green appearance, compared with cool-season species.

Flint (1903) was observant of characteristics of warm-season grasses, and the lack of suitable types in the southeastern USA. Possibly none of the major sod-forming turf species that we currently grow were native to the southeastern USA. The first turf immigrants may have been stowaways. Evidence for this is circumstantial in the case of bermudagrass, but convincing,

> The earliest herbarium specimens and records in the United States are from a string of naval stations and ports along the East Coast. [Bermudagrass] had repeatedly been introduced with ship's ballast, and a good many other weeds found their way to our shores the same way. (Harlan, 1975).

Sauer (1972) ruled out seed flotation as a means of transoceanic dispersal of St. Augustinegrass, but he also placed its distribution as part of a "larger migrational mystery involving . . . other cosmopolitan tropical seashore grasses that lack proven capability of long-range sea dispersal."

During the early 1900s, there were extraordinary efforts of public and private plant collectors, who introduced or augmented the germplasm that is now available to plant breeders. The grasses came from Africa (bermudagrasses, *Cynodon* spp.); the West Indies [possibly carpetgrass, *Axonopus affinis* Chase and St. Augustinegrass, *Stenotaphrum secundatum* (Walt.) Kuntze, either of which may have been native to the southeastern United States]; East Asia (zoysiagrasses, *Zoysia* spp.; and centipedegrass, *Eremochloa ophiuroides* (Munro) Hack.]; and South America (bahiagrass, *Paspalum notatum* Flügge).

The need for conservation grasses was great. Thin, erodible soils needed protection from the heavy, erratic rains. The early (pre-1940s) public references to present turf species were primarily in the context of forage and soil conservation. The final selection and naming of turf types was done by the private sector, e.g., golf course superintendents and sod producers. The early years provided us with several cultivars, but few references, and little living record of the plants used. In later years of warm-season turfgrass breeding the number of references increases, and considerable effort occurs in resolving basic breeding problems (e.g., mode of reproduction, population improvement, pest resistance, and climatic adaptation). Industry also developed a renewed interest in participating in cultivar development (White & Busey, 1987).

Plant breeding achievements usually have a direct effect on a crop (e.g., improved yield). There is also an interactive, "niching" effect (e.g., dwarf cereal cultivars have less lodging problems at high fertility). In warm-season turfgrasses, improved cultivars have a large interaction with improved management. Improved cultivars (e.g., 'Tifdwarf' bermudagrass) often depend upon specialized management for them to reach adequate turf consistency. Plant breeders must consider the management environment in developing new cultivars. In the event that management can readily deal with a turf problem, the breeder often does not attempt to find a genetic solution. Similarly, a turfgrass cultivar might be selected because it is particularly well suited to a management condition, e.g., a turfgrass must be resistant to a commonly used herbicide. The genotype-management synergism benefits humanity by providing a more utilitarian turf product, which could not be achieved by genotype or management alone. It also stimulates a viable turf industry which employs thousands.

ERAGROSTOIDEAE: CHLORIDEAE

The Eragrostoideae are C_4 grasses that are more often found in warm, dry climates (Gould, 1968). They include some underdeveloped species, e.g., buffalograss [*Buchloë dactyloides* (Nutt.) Englem.] that provide utilitarian low-maintenance turf in the southern Great Plains. 'Texoka' (Voigt et al., 1975) is one widely distributed cultivar. Research has been done to unravel the factors involved with sex expression in this species (Huff & Wu, 1987)

Bermudagrass

Origin and Early History

Bermudagrass (*Cynodon* spp.) was introduced to the USA from Africa by 1751 (Hanson, 1972a). The high growth rate of this genus (Busey & Myers, 1979) provides it with rapid colonization in disturbed areas and quick recovery from traffic damage caused in sports activities. It was shown that if logarithmic growth rates could be sustained for 1 yr (e.g., through ideal environment and frequent division), then "it would be possible for 1 m^2 of grass to cover an area equal to 50% of the land area of the world" (Busey & Myers, 1979).

Carolina farmers of the 1700s preferred it as a forage, calling it "crop grass or crab grass (*Syntherisma*)", according to Drayton's *View of South Carolina* (Gray, 1958). Bermudagrass was planted on golf courses (Carrier, 1927) and lawns (Enlow & Stokes, 1929) in Florida at least by the 1920s. 'St. Lucie' was the first turf bermudagrass recognized in the USA (Tracy, 1917). St. Lucie was a slender dwarf plant used in Florida lawns, but it was not hardy as far north as middle Georgia. The fine-leaved texture of bermudagrass was recognized as an advantage for use in lawns, but its intolerance to shade was also noted (Enlow & Stokes, 1929). Arizona-grown (probably common, *C. dactylon*) seed was planted in most golf course fairways, greens, and tees in Florida in the 1940s (R.A. Bair, 1987, personal communication).

Species and Cultivars

Eight species of *Cynodon* are recognized, of which turf types are included only in the $2n = 36$ *C. dactylon* (L.) Pers. var. *dactylon* (common bermudagrass), the $2n = 18$ *C. transvaalensis* Burtt-Davy (African bermudagrass), and their $2n = 27$ interspecific hybrid *C.* × *magenissii* Hurcombe (= *C. dactylon* × *C. transvaalensis*) (Harlan et al., 1970a). My interpretation of *C. magenissii* being equivalent to *C. dactylon* × *C. transvaalensis,* and reciprocals, has not been recognized by taxonomists, but should be studied. This tri-partite interpretation of turf species relationships in *Cynodon* is a simplified classification based on incomplete knowledge. Extensive studies of genetic variation have emphasized wide crosses (Harlan & de Wet, 1969) and forage development, but little documentation is available regarding the taxonomy and relationships of cultivated turf types. Cytological evidence suggests that *C. transvaalensis* could be considered a botanical variety of *C. dactylon,* but that it is distinctive in geography, ecology, and morphology (Harlan et al., 1970b). Taxonomy of bermudagrass is helpful in understanding its breeding development.

Early selection work emphasized *C. dactylon*. Probably much of the germplasm base for early selections was from seeded plantings. At one time all the commercial seed used in the USA came from Australia (Tracy, 1917). The first test plantings were reported to have been made around 1918 at East Lake Country Club by C.V. Piper of the U.S. Golf Association Green Section. Out of this came the 'Atlanta' strain about 1924 (Latham, 1966). Additional clones were selected by golf course superintendents in the USA and the Republic of South Africa. There was little organized breeding. Important cultivars included 'Ormond', by R.A. Bair, released in Florida in 1962; 'Royal Cape', by C.M. Murray, Republic of South Africa, in 1930; and 'Tiflawn' (T-57), by G.W. Burton, Georgia, distributed in 1952 (officially released 1956). Ormond and Tiflawn have had a continuing success for use in recreational turf. Tiflawn tolerates heavy traffic, requires little fertilization, is fast spreading, and is generally disease and insect resistant (Hanson, 1972b). Ormond is blue-green, competitive, and performs moderately well under conditions of natural pest infestation (*Scapteriscus* mole crickets, especially) and suboptimal N fertilization (Busey, 1986a).

The foregoing cultivars are sometimes lumped together as common bermudagrass. The name "common bermudagrass" is confusing and questionable. In the USA, it refers to a landrace cultivar that is produced as seed in Arizona and other states, is widely recognized, but has not been adequately described. When used as a collective term, "common" can also refer to any *C. dactylon.* In some usages, common has become synonymous with any seed propagated bermudagrass. Because of its coarse habit of growth, there is relatively little interest in using plants of this group in the subtropics, where some of the sterile hybrids (below) can be used. In temperate areas where cold can be a problem, *C. dactylon* is better adapted than other *Cynodon* species, and there has been a continuing effort to improve cold hardiness and turf characteristics. In 1930, A.B. Dorrance started to select bermudagrass for cold hardiness in Augusta, MI (Hanson, 1972a). 'Midiron', developed by R.A. Keen of Kansas State Univ., shows considerably less winter-kill than several other cultivars (Juska & Murray, 1974). Burton (1974) produced hundreds of F_1 hybrids involving a cold-hardy Berlin collection. The cv. Vamont was released from Virginia for its cold hardiness. Wofford and Baltensperger (1985) performed extensive heritability tests for turfgrass characteristics in *C. dactylon.*

Clones of the $2n = 18$ *C. transvaalensis,* including 'Uganda', were introduced to the USA. They crossed with the naturalized $2n = 36$ *C. dacty-*

lon to produce sterile $2n = 27$ hybrids. The products of these crosses became useful cultivars, such as 'Sunturf', 'Everglades-1', 'Gene Tift' (='Bayshore'), and a long line of successful Tifton hybrids. In general turf usage, "hybrid" bermudagrass has come to mean any interspecific hybrid involving *C. dactylon* and *C. transvaalensis*. The first Tifton interspecific hybrid was 'Tiffine'. Tiffine was a cross of *C. transvaalensis* × *C. dactylon* (Forbes & Burton, 1963). It was too stemmy for use on putting greens (Latham, 1966).

A remarkably successful hybrid cv. Tifgreen (Tifton-328, *C. dactylon* × *C. transvaalensis*) was officially released in 1956, and it made an excellent putting surface. It was followed by 'Tifway' (Tifton-419, *C. transvaalensis* × *C. dactylon*) in 1960. Tifway had a stiffer, more erect growth habit, and greater pest and wear resistance than Tifgreen. In 1974, Tifgreen and Tifway represented 78% of the areas planted to specific grasses on Florida golf course greens, tees, and fairways (Anonymous, 1976). One must be cautious in accepting that estimate, which was based on interviews of turfgrass managers, because I have observed that some areas of golf course fairways and sod fields that are claimed to be Tifway are actually Ormond (P. Busey, 1985, personal observation). The leaf blades of Tifway are pilose, with 10 or more hairs on both surfaces, whereas those of Ormond are sparingly pilose (Thompson, 1965).

Clonal Variation

Clonal offtypes were discovered in Tifgreen, and they were presumed to have arisen as spontaneous mutations. Two offtypes were selected as cultivars, Tifdwarf (USDA Coastal Plains Experiment Station, Tifton, GA, 1965) and Pee Dee 102 (South Carolina Agric. Exp. Stn., 1968). Under close cutting, Tifdwarf can provide rapid golf ball roll. The problem of vegetative sporting in hybrid bermudagrass is widespread, but has not been adequately documented. Because of the rare nature of mutation events, it is not likely that one could prove that mutations were involved, to the exclusion of alternative hypotheses (e.g., contamination of breeding stock). I have been shown several instances of offtypes that had been observed during the growing in of planting stock. This suggests that the immediate problem may be contamination of planting stock, even though mutations ultimately may have been involved.

Maintenance of genetic purity is a major problem for golf course superintendents, turfgrass producers, and plant breeders. "Clonal degeneration" is not an isolated phenomenon in bermudagrass, but has been noticed in potato (*Solanum tuberosum* L.) and other crops, in which a "running out" may occur. Several different phenomenon can occur, including the "risk of multiplying mutations or cryptically diseased material" (Simmonds, 1972). Clonal selection also allows for openings of genetic vulnerability, because of the fundamental genetic similarity of clonal offtypes.

Golf course superintendents have continued to make new vegetative selections from the offtypes that they find. This will be a viable source of variability in quality characteristics. A synthetic approach to this goal has been the use of non-ionizing radiation, e.g., gamma rays, to create artificial mutations. Beginning in 1970, Powell et al. (1974) irradiated thousands of rhizomes of Tifton bermudagrass cultivars. Two clonal cultivars were released from this work, 'Tifway-II' in 1981 (Burton, 1985) and 'Tifgreen-II' in 1983. Tissue culture techniques may now be useful for increasing bermudagrass clonal variations, because plant regeneration has been shown to occur from calli (Ahn et al., 1985).

Environmental Stresses

There is considerable undocumented field information on pest resistance of bermudagrass cultivars, which has served as a basis for cultivar descriptions and releases. The few published studies have been based on individual experiments involving a small number of bermudagrasses grown in containers. Those experiments have not been adequately correlated with field studies. One might speculate (Busey, 1986a) that the low-maintenance adaptation of Ormond and some plant introductions (e.g., PI 291586) is related to pest resistance, i.e., *Scapteriscus* mole crickets (Reinert & Busey, 1984), ectoparasitic nematodes (Tarjan & Busey, 1985), tropical sod webworm, *Herpetogramma phaeopteralis* Guenée (Reinert & Busey, 1983), and bermudagrass mite, *Eriophyes cynodoniensis* Sayed (Reinert et al., 1978). Unfortunately, it is difficult to extrapolate to field level resistance without more information. For highly mobile grass pests such as caterpillars (lepidopteran larvae), cafeteria-style grazing trials (Leuck et al., 1968; Reinert & Busey, 1983) are of undetermined value. Insects in the field are generally confronted with turfgrass monocultures, and the element of nonpreference which is noticeable in controlled screening trials would be of little or no impact in the field, unless it is correlated with other resistance mechanisms.

Physiological studies were done to select for better adaptive range for bermudagrass. 'No-Mow' (FB-137) bermudagrass is more tolerant to shade than 'Tifway', based on color and ground cover ratings when both grasses are grown at 35 and 25% incident light (McBee & Holt, 1966). 'Brookings' bermudagrass has significantly higher regrowth than seven other cultivars, when all are subjected to -4 to $-11\,°C$ soil temperatures (Ibitayo et al., 1981). Electrolyte loss curve data suggested that the low temperature kill point of hardened material ranged from -7 to $-17\,°C$, depending on the bermudagrass cultivar. Salt-tolerance differences exist among bermudagrasses (Dudeck et al., 1983), with Tifdwarf showing maximum root growth at moderately high (20.5 dS m^{-1}) electrical conductivity (EC), or about 47% the concentration of seawater. Other cultivars show maximum root growth at lower EC values. One of the advantages of bermudagrass is its high N responsiveness, which provides a recuperative ability against wear and herbivory. Yet some work has been done in Florida, New Mexico, and other locations, to select for adapted, low-fertility requiring bermudagrasses. Differences in air pollution (ozone and peroxyacetyl nitrate) tolerance exist among bermudagrasses, with Tifgreen being intermediate in response, while Santa Ana and Common are not injured (Youngner & Nudge, 1980).

ERAGROSTOIDEAE: ZOYSIEAE

Zoysiagrasses

Zoysiagrasses (*Zoysia* spp.) are naturally distributed (Ohwi, 1965) and grown extensively as a turf (Maki, 1976) in temperate east Asia, and two species extend throughout Malesia (Goudswaard, 1980).

Unlike other warm-season grasses, zoysiagrasses flower in response to short days (Youngner, 1961). Because of their high density and rhizomatous growth habit, zoysiagrasses tend to crowd out other plants.

Zoysia was probably involved in the first recorded reference to turf, in the "Man-yo-shu", a collection of poems from 759 A.D. (Maki, 1976). Subsequent references to sodding occurred in the *Saku-tei-ki,* a gardening book published in Japan in 1156. According to Maki (1976), the lawn garden was connected with the tea ceremony, and commercial sod farming began in 1701. *Zoysia japonica* Steud. and other species grow wild in native grasslands in Japan. A finer species, *Z. matrella* (L.) Merr., was introduced to Japan from China in 1713. In the late 1920s, "Koraishiba" (*Zoysia* spp.) was used on the course at the Kobe Golf Club in Japan (Kakuda, 1987). By 1970, there were 26 cultivars of *Zoysia* reported in Japan (Kitamura, 1970; cited by Maki, 1976).

Zoysiagrass was introduced to the USA before 1895 (Hanson et al., 1974). In 1902, David Fairchild (Fairchild, 1938) observed "Birodoshiba" (*Z. pungens* Willd. = *Z. matrella*) growing as a velvet-like turf in gardens around Tokyo. He introduced PI 9299, which may have been the first recorded zoysiagrass in the USA. In 1912, Piper observed attractive lawns of zoysiagrass in the Philippines (Piper, 1915). The value of Zoysia in the USA was immediately obvious when it was introduced. By 1916, it was stated that zoysiagrass "seems to be succeeding in Florida as a lawn grass" and " is worthy of trial in sand-hill districts or on saline lands near the coast" (Fairchild, 1920). *Zoysia* has been used principally in the warm- and cool-season transitional belt from New England to Missouri and further south.

Clonal selections from the USA have included *Z. japonica* Steud. 'Meyer' (1951, USDA and U.S. Golf Assoc.), which is adapted to the north Atlantic coast of the USA, and *Z. japonica* × *Z. matrella* var. *pacifica* Goudswaard (= *Z. tenuifolia* Thiele = *Z. tenuifolia* Willd. ex Trin.) 'Emerald' (Forbes, 1962). Because of the extensive cross-compatibility of zoysia species (Forbes, 1952), they may be considered as one species (I. Forbes, 1976, personal communication). *Zoysia matrella* is grown in south Asia (Greenfield, 1974) but establishes slowly in Kenya (Bogdan, 1970). In addition to its slow establishment, major problems with zoysiagrasses include excessive thatch development (Dunn et al., 1981), and nematodes (Busey et al., 1982b; Grisham et al., 1974).

There has been renewed interest in the use of zoysia in low-maintenance areas, for example, roadsides, where it might be intermixed with other species, e.g., tall fescue (*Festuca arundinacea* Schreb.) (J.J. Murray, personal communication, 1987). Strains of *Z. japonica* or possibly *Z. micrantha* from Korea have shown promise for establishment from seed.

PANICOIDEAE: ANDROPOGONEAE

The Panicoideae are C_4 grasses that are "particularly successful in moist, humid, tropical, or subtropical habitats" (Gould, 1968). They represent the largest diversity of turf genera and species for use in warm-season areas. There are a number of underdeveloped species, some of which serve people without much formal recognition. One of these, hurricanegrass or Seymourgrass, *Bothriochloa pertusa* (L.) A. Camus, is commonly mowed in the West Indies, and forms a dependable, low density turfgrass.

Centipedegrass

Centipedegrass [*Eremochloa ophiuroides* (Munro) Hack.], was first brought to the USA from China in 1918, in the baggage of plant explorer Frank N. Meyer, who had been collecting in Hunan Province (Fairchild, 1938). Meyer was missing from the steamer in which he had been traveling, and it was presumed that he fell overboard in the Yangtze River. By 1925, centipedegrass was grown at the Belle Glade Exp. Stn. (later known as the Everglades Exp. Stn.) under the name "Chinese love grass" (see herbarium specimen collected by the Rev. O'Neill, and accessioned at Missouri Botanical Garden Herbarium).

Centipedegrass is an excellent turf for heavier soils of low fertility; N application was shown to reduce centipedegrass appearance (Walker & Burns, 1975) and increase killing temperature from about -12 to $-9\,°C$ (Palmertree et al., 1974). The cultivar 'Oklawn' was released in 1965 by the Oklahoma Agric. Exp. Stn. as a cold hardy type. 'Tennessee Hardy' and several numbered selections were shown to be comparably cold hardy (Johnson & Dickens, 1977).

Centipedegrass is a sexually reproducing diploid ($2n = 18$) which has some degree of self-incompatibility (Hanna & Burton, 1978). The lack of genetic variation in the species has been noted (Bouton et al., 1983). This is explained by the fact that before 1974, only 12 accessions of centipedegrass were recorded in the USA (Anonymous, 1903–1974). Additional introduction work has been concerned with a broader sampling of clones from China (Hanna & Burton, 1978) including the wild relative *E. ciliaris* (L.) Merr. (Hanna & Shu-Liang, 1986). Although the latter is a bunchgrass and not winter-hardy at Tifton, GA, some progeny were observed to fail to flower. This suggested a short-daylength sensitivity of *E. ciliaris* and a possible use as a germplasm from which to produce a nonflowering *E. ophiuroides,* if the two species can be crossed (Hanna & Shu-Liang, 1986).

Mutation breeding has been performed on centipedegrass seed (Johnson et al., 1977) with interest to obtain variants for morphological traits and cold hardiness. Out of approximately 8000 seedlings that were checked, the cv. AU Centennial was selected in 1983 (Pedersen & Dickens, 1985). Compared with common centipedegrass, AU Centennial is denser, finer textured, and darker green.

PANICOIDEAE: PANICEAE

Carpetgrass

Relatives of *Axonopus* (below) such as *Digitaria* are useful for turf, and worthy of further development. One species, *D. didactyla* Willd., appears to have promise (H.F. Whiting, ca. 1984 personal communication). For taller growing turf, *Pennisetum ciliare* (L.) Link has been used. Kikuyugrass, *P. cladestinum* Hochst. ex Chiov., is important in California. Species of *Panicum,* such as *P. ramosum* L., browntop millet, are used as companion grasses for plantings of bahiagrass, but this practice is not recommended, because the millet is too competitive against the bahiagrass (P. Busey, 1987, unpublished data).

Salt-tolerant siltgrass *Paspalum distichum* L. (= *P. vaginatum* Sw.) (Fosberg, 1977) is useful for golf courses and other turf areas that receive brackish water. The cv. Adelayd, selected by H.F. Whiting, has superior density compared with wild types. Hydroponic studies show that Adelayd may not be as salt tolerant as other selections, such as FSP-1 (Dudeck & Peacock, 1985). Morton (1973) summarized much of the early history of silt grass.

Carpetgrass (*Axonopus affinis* Chase) is considered a native of the southeastern USA (Chase, 1938) but may have been introduced from the West Indies (Burton, 1951). It grows best in moist soil (Martin, 1975) and in Florida it occurs occasionally in swales and under trees. It also grows in nonirrigated areas in Louisiana, where soils are heavier. In sand soil, carpetgrass is less drought resistant than either bermudagrass or bahiagrass, and has a relatively shallow root system (Burton et al., 1954). Carpetgrass tolerates moderate shade. It is available as seed. No cultivars are known and no breeding work appears to have been done. Although once considered the most common pasture grass in Florida (Burton, 1951), carpetgrass has been largely replaced by more nutritious species.

Carpetgrass is potentially advantageous for low-lying areas, especially where soil fertility is low. It shares its mat-forming stoloniferous growth habit with closely related genera such as *Brachiaria* and *Digitaria.* In some of the Windward Islands (such as Martinique) carpetgrass grows well at moderate elevations, where there is evidently sufficient cloud cover and regular precipitation (P. Busey, 1980, personal observation). There are other species, such as *A. compressus* (Swartz) Beauv. which have turf characteristics but broader leaves.

Bahiagrass

Bahiagrass (*Paspalum notatum* Flügge) was not mentioned as a turf in 1929 by Enlow and Stokes (1929) and received little research attention until the late 1930s. In 1938 Escambia, Florida County Agent E.H. Finlayson discovered what would be called 'Pensacola' bahiagrass (*Paspalum notatum* Flügge var. *saurae* Parodi) (Parodi, 1948) on a sodded sand bank (Finlayson, 1941). This excellent-growing plant may have arrived as a stowaway on a fruit boat from Central or South America (Hanson, 1972b). It was released in 1944. Burton (1946) discovered Pensacola is diploid ($2n = 20$) and reproduces sexually based on variable progenies (Burton, 1955). Native to eastern Argentina (Quarin et al., 1984), the Pensacola cytotype has become one of the major forage grasses of the southeastern USA (Burton, 1967). It is planted extensively from seed along highways in Florida, North Carolina, and other subtropical and mild temperate areas. Pensacola is more cold tolerant than tetraploid bahiagrass cultivars.

The tetraploid bahiagrass cytotype *P. notatum* var. *latiflorum* Doell ($2n = 40$) is the most common botanical variety in tropical and subtropical America (Parodi, 1948). It has considerable variability (Fernandes et al., 1973) and includes several broad-leaved cultivars, such as 'Argentine', 'Paraguay-22', and 'Wilmington'. The former is widely used in lawns and for solid sodding on highway embankments and the 0.8-m edge along the pavement. Argentine has broader leaf blades than Pensacola, yet it is superior as a turf because of its more prostrate, denser-appearing growth habit. This violates the usual precept that finer-textured cultivars are superior turf types. Argentine was released in 1950 by the Univ. of Florida as a pasture grass (Killinger et al., 1951). It had been introduced from Argentina in 1944 as PI 148996. Naturally occurring, tetraploid bahiagrasses are always obligate apomicts (Burton, 1948). Breeding improvement has to be either clonal selection, or else must cross the reproductive barrier by the use of artificially created tetraploids which are sexual (Burton & Forbes, 1960). The F_2's from sexual × apomictic crosses shows that apomixis in tetraploid bahiagrass is mostly recessive.

The main turf use for bahiagrass is for large turf areas. Soil erosion is considerably reduced. In Taiwan continuous bahiagrass cover reduces annual soil loss from $1.56 \cdot 10^4$ g m^{-2} (with clean culture) to $1.5 \cdot 10^2$ g m^{-2} (Jean & Juang, 1979). Bahiagrass turf requires little or no irrigation, even in well-drained sands, requires minimal fertilization, 5 to 10 g of N m^{-2} yr^{-1}, and has few important pest problems, e.g., *Scapteriscus* mole crickets. The use of bahiagrass for turf has often been discounted, yet in 1974 it was the principal turfgrass of Florida (Anonymous, 1976). Unfortunately, bahiagrass grows poorly in areas with moderate or heavy shade. It produces abundant seedheads for a brief period (mostly about 6 wk in summer), but some seedheads occur sporadically throughout the growing season. Bahiagrass managers recognize the species to be difficult to keep mowed. The seedheads typically reach about 0.7-m height.

Seed production is meager (225–350 kg ha^{-1} would be the high range, W.R. Cook, 1982, personal communication) and seedling establishment is often poor. Because of its slow vegetative growth rate (Busey & Myers, 1979) and open growth habit, bahiagrass is susceptible to weed invasion, particularly in the seedling stage. Better management can correct establishment problems of bahiagrass. A Rapid Coverage Polycross (RCP-1) population was developed in Florida for possible cultivar release (P. Busey, 1983–1987, unpublished data).

The seedhead problem has been evaluated for possible reduction through selection. Efforts have been made to develop a dwarf, turf-type bahiagrass

(W.W. Hanna & G.W. Burton, 1987, personal communications; Busey, 1985). No selection appears to have been done towards lessening the shade sensitivity of bahiagrass. *Scapteriscus* mole crickets screening yielded inconclusive results (P. Busey & J.A. Reinert, 1978, 1983 unpublished data). This pest damages turf mechanically while foraging, much as a rototiller would damage turf, and in some cases the turf is not the host. Thus, it may be unfeasible to develop mole cricket resistant cultivars. The majority of research on bahiagrass improvement continues to be for forage. This includes work in Japan (Sakai, 1983), the USA (Burton, 1982), and Zimbabwe (Mills & Boultwood, 1978).

Bahiagrass cultivars have not been developed specifically for turf, yet the recognition of cultivars for pasture and conservation has satisfied some turf needs. A major concern in the development of any turf-type bahiagrass is whether seed producers will accept it. Bahiagrass seed is traditionally a by-product of the cattle industry. The production of seed of a new cultivar on any existing bahiagrass area would entail a major difficulty in the eradication of existing grass. An improved turf bahiagrass, to be successsful, would need either to be competitive and useful as a forage, or it would need a potential market sufficiently large to justify specialized turf seed production areas.

St. Augustinegrass

St. Augustinegrass [*Stenotaphrum secundatum* (Walt.) Kuntze], had an early distribution in North America, even though the other six species of the genus are strictly Old World (Sauer, 1972). Its closest wild relative, pembagrass, *S. dimidiatum* (L.) Brong., is sometimes used in east Africa for turf (Sauer, 1972). Some writers have considered St. Augustinegrass to have come from the West Indies. Nevertheless, I have not seen plants from the West Indies like our naturalized Gulf Coast Group (Busey et al., 1982a), either living or as herbarium specimens. The type specimen of *S. secundatum* was collected in the Carolinas in the 1700s, which is consistent with several alternative hypotheses regarding its introduction, or even native status, in North America.

The earliest recorded use of St. Augustinegrass as turf was in 1880, when it was planted along an avenue at A.M. Reed's plantation near Orange Park, FL (unpublished diary of A.M. Reed and J.P. Waterbury, 1987, personal communication). There were more frequent collections of St. Augustinegrass in the 1880s and 1890s from Louisiana and Florida, including nonport areas (e.g., Florida central highlands). It is likely that during that era the species served as a forage. It grows in remote areas in Everglades National Park, e.g., Highland Beach (P. Busey, 1987 unpublished observation) which was inhabited briefly in the 1890s by people with cattle.

Before 1920, there were St. Augustinegrass lawns in Davie, FL and by the late 1920s sod trucks were observed in Miami (R.R. Hammer, 1984, personal communication). The common St. Augustinegrass of that era was not marked by a particular name. Based on plants maintained in breeding collections, a survey of existing lawns (Busey, 1986b), and a study of herbarium specimens (P. Busey, 1978–1985, unpublished data) there were at least two "common" types. 'Florida Common' had long, purplish internodes on the stolons and purple stigmas, whereas 'Texas Common' of the Gulf Coast Group had green internodes and white stigmas. St. Augustinegrass sod production in Florida was a matter of stripping out pastures so there was initially little incentive to plant improved cultivars.

In the mid-1930s, several sod growers (White & Busey, 1987) introduced the cv. Bitterblue. Although incorrectly believed to be resistant to the southern chinch bug (*Blissus insularis* Barber) Bitterblue was superior to Florida Common in other respects. Bitterblue had a deeper blue-green blade color, denser growth, and shorter height. Bitterblue is susceptible to gray leaf spot disease caused by *Pyricularia grisea* (Cke.) Sacc. (Atilano & Busey, 1983), which is still a problem in the production of that cultivar. A pasture grass, 'Roselawn', was distributed in 1942 and 1943 by Everglades Exp. Stn., Univ. of Florida (Allen & Kidder, 1971). After much Bitterblue was destroyed in Davie, by the hurricane of 1947, ranchers around Lake Okeechobee began to cut Roselawn pastures and sell the product as sod. As with Florida common, Roselawn has an open habit of growth, is light colored, and does not form a dense sod.

Concerns about variety purity led the Univ. of Florida to release a Bitterblue-like plant that was called 'Floratine' (Nutter & Allen, 1960). Meanwhile, Florida sod producers had a state-sponsored certification program implemented. Quality assurance was helpful in marketing, by providing the consumer a dependable product. It also protected against the spread of pest organisms. In all warm-season turfgrasses, there are unresolved questions of cultivar identity, and assurance of cultivar purity is essential for successful breeding development and cultivar adoption to occur.

In 1973, The Univ. of Florida and Texas A&M University released 'Floratam' (Horn et al., 1973), which is resistant to the southern chinch bug (Reinert & Dudeck, 1974) and to St. Augustine Decline Strain of Panicum Mosaic Virus (PMV-SAD). The Floratam cultivar spread and became so popular that by 1980–81 it dominated 77% of commercial sod samples in southeast Florida, and had penetrated 21% of lawn areas (Busey, 1986b).

All the PMV-SAD resistant genotypes, with one exception, are clones with purple stigmas and purple stolons (Bruton et al., 1983). Likewise, only polyploid St. Augustinegrasses have antibiotic resistance to the southern chinch bug (Reinert et al., 1986), although in some cases the southern chinch bug has adapted to reproduce on and kill Floratam (Busey & Center, 1987).

O.M. Scotts and Sons contributed significantly towards the development of dwarf St. Augustinegrass cultivars. Breeding work by T.P. Riordan and others (Riordan et al., 1980) resulted in the development of 'Seville' (= S-6-68-516; Plant Patent 4097) and other strains that are under commercial development ('Delmar' = S-6-72-99 and 'Jade' = S-6-72-182). As a group, these plants are more susceptible to drought and to the southern chinch bug than Floratam. They are also more shade tolerant, lower growing, and more aesthetically acceptable to most observers than Floratam.

Seedhead production is a major problem of some diploid ($2n = 18$) St. Augustinegrasses, and this can markedly reduce the appearance value of otherwise acceptable Seville, particularly during warm summer months. Yet there is one strain, Plant Patent 3834 by C.L. Garrett, which has been developed as a seed-propagated type. Seedhead production in St. Augustinegrass is under photoperiod control, but there are some genotypic differences in the dependence on this stimulus (Dudeck, 1974).

Severe winter-kill occurs to St. Augustinegrass following temperatures of −5 to −10 °C, although in a cold resistance study in Mississippi, temperatures as low as −15 °C were experienced (Wilson et al., 1977). Floratam has no survival from such cold temperatures, whereas 'Raleigh' has 100% survival. Raleigh, developed by W.B. Gilbert, North Carolina State Univ., was released for "commercial trial" in 1980. Another advantage of Raleigh is its resistance to PMV-SAD (Reinert et al., 1980). This was an exception to the association of purple stigmas with PMV-SAD resistance, since Raleigh has white stigmas (Busey et al., 1982a). St. Augustinegrass cold damage is related to management, e.g., excessive N fertilization (12 g of N m^{-2}) applied in November and December (Reeves & McBee, 1972), but the genotype factor is also great. Sod producers in Texas and northern Florida experienced considerable loss of stand in Floratam due to cold weather, and converted many areas to Raleigh and Texas Common, which was also relatively cold hardy. Both Texas Common and Raleigh are in the Gulf Coast Group, which is apparently endemic to the USA (Busey et al., 1982a).

Mutation breeding was shown to create favorable variations of St. Augustinegrass (Busey, 1980; Powell & Toler, 1980) while preserving resistance characteristics (Reinert et al., 1981). No cultivars have resulted from this work.

Inconclusive results have been obtained from attempts to measure shade tolerance in St. Augustinegrass (Peacock & Dudeck, 1981; Winstead & Ward, 1974), although Floratam has been shown to be quite intolerant of shade (Barrios et al., 1986). The general progress of understanding mechanisms of environmental adaptation in St. Augustinegrass has been slow, but practical results (e.g., shade tolerance in the Scotts germplasm, and cold tolerance in the Gulf Coast Group) have been encouraging. These associations, along with similar racial and/or ploidy generalizations for gray leaf spot disease (Atilano & Busey, 1983), lepidopteran larvae damage (Busey et al., 1982a), and chinch bug resistance (Reinert et al., 1986), support the idea that "classification should serve as a prospectus of beneficial genes" (Busey et al., 1982a). Recently it was discovered (P. Busey, 1988, unpublished data) that certain polyploid introductions from Africa have the ability to survive greatly extended dry periods (months) of no irrigation, provided relatively shallow (1.4 m) groundwater is present. For coastal areas where St. Augustinegrass is commonly used, such a germplasm may have tremendous value.

DISCUSSION AND CONCLUSION

These highlights have demonstrated that much progress and many benefits have been gained from warm-season turfgrass improvement, but that much more must be done.

Modern cultivars are denser and more traffic tolerant than the common types. This provides safety and sanitation. Modern cultivars are lower growing and have narrower leaf blades. This provides a higher aesthetic beauty. Modern cultivars are in some instances resistant to important risks, such as pests. In some four decades of activity, this progress in warm-season turfgrasses has been accomplished through the part-time activity by a few dedicated persons. More recently, since 1975, there have been one or at most two scientist positions dedicated full-time to the effort, in addition to the continuing support by cooperators, both public and private. What progress can we expect to see, and how can it be accomplished?

Recurrent Selection

Recurrent selection involving generations of hybridization has been conducted infrequently in warm-season turfgrasses. Past efforts that have lacked the recurrent approach have been sporadic, but successful. Most early success have been the introduction of new species, or the serendipitous discovery of racial ecotypes, interspecific hybrids, or naturally occurring mutations. While this is turfgrass breeding, it has not adequately established a platform for further advance. Recurrent selection must be conducted for much further progress to be made. This depends, in turn, on an appreciation of what has been already learned, and what research is still needed.

Where recurrent selection is carried out, as it must be, the taxonomy and breeding method of warm-season grass species must be resolved. In addition, known breeding stocks must be described and their diversity preserved, because future progress can be limited by our present (or future) starting point. This will require much more effort and care than has been applied in the past. Today we are not even secure about the identity of major cultivars. In view of the problems of genetic purity, much more work should be done to maintain collections and to characterize them. This is particularly noticeable because of the emphasis on clonal selection.

Clonal Selection

Virtually all progress in warm-season turfgrass breeding involves clonal selection. Vegetative propagation is necessary in certain ploidy conditions. Seedhead production, where it occurs abundantly, is a severe aesthetic flaw. Clonal selection has been used even in normally seed propagated warm-season grasses, whether as a result of apomixis (e.g., Argentine bahiagrass) or because of the need for uniformity (e.g., AU Centennial centipedegrass). Clonal selection allows for the accurate recognition of superior genotypes, and it facilitates their expansion and distribution as new cultivars. While clonal selection has been an obvious opportunity for the breeder, it has entailed liabilities that have hindered recurrent selection.

Clonal germplasm is difficult to maintain, thus parent lines of important warm-season turfgrass cultivars have in most cases vanished. A dependence on seed production in the course of hybridization may naturally select for seedy types, unless inflorescence development is controlled in some manner. Errant seeds that germinate are a source of offtypes that may noticeably contaminate a vegetatively propagated cultivar. These dilemmas require continued research on seedhead initiation and fertility in warm-season turfgrasses, even though a commitment to clonal selection has been made.

Artificial mutation breeding is an appealing approach for the stepwise improvement of warm-season grasses, but it has been tried repeatedly with few successes. Naturally occurring mutations have most probably been involved in the origin of many cultivars, and there should be continued opportunity to use this source of variation. The approach has been successful in other vegetatively propagated crops, albeit mostly in cases where prominent color variations are marketed. There are difficulties in recognizing complex adaptive variations from mutations; sometimes a very large number of genotypes must be winnowed. Neither naturally nor artificially occurring mutations do much to broaden the genetic base.

As an added caution, vegetatively propagated crops are prone to the problems of offtypes, and some germplasm is more prone to continued natural mutations, both beneficial and detrimental. If mutability is a genetic trait in warm-season turfgrasses, then we could infer that selection of mutants would entail the risk of selecting for mutability, in the process of mutation breeding. Perhaps this could partly explain the frequency of offtypes that occurs in Tifdwarf bermudagrass. Regardless of which approach (hybridization or mutation) is used to diversify germplasm, more research is needed on mutations and genetic variations. If this is not done, whatever progress is made in clonal selection will be dissolved into confusion and contamination by offtypes.

Selection Criteria and Environment

Much of the success of warm-season turfgrass breeding is due to direct field selection, even though the adaptation of warm-season turfgrasses is only partially understood. Direct selection can be successful, without attention to its mechanism or basis. Despite the best laid plans for an array of resistance criteria and other indirect selection criteria, plant breeders often pick plants that grow well, and it is years later that plant physiologists and other specialists discover why. There are a few instances of documented pest resistance, and many observations of environmental stress resistance. These examples of "indirect selection" have been used mostly in description, occasionally in final screening, and rarely in germplasm development. Although field-oriented selection may appear risky, statistical genetics proves that indirect methods are often even less likely to result in genetic advance.

A general proviso for either indirect or direct selection to succeed is that sufficient genetic variance must be present. The added conditions specific to indirect selection are (i) that a trait be accurately recognizable in a pot, a laboratory, a growth chamber, etc. (and thus the trait in that environment has a low error variance), and (ii) that the indirect measure be highly correlated with a field response. As an example, resistance to an acutely damaging pest might be easily recognizable and correlated between the greenhouse and the field, and thus effectively selected in the greenhouse.

Resistance to an acutely damaging pest might fulfill another desirable condition of indirect selection, which is that the pest problem be serious enough to warrant a special effort. Finally, a justification for indirect selection criteria is that such a procedure saves the plant breeder time. This would be most supportable in the case of a trait (e.g., thatch development or nematodes) that is not expressed early in the field evaluation, but would be partially refuted if it required a whole research project to develop the indirect selection criteria into a usable tool.

Based on the foregoing conditions for indirect selection, it is likely that the field approach is better in most instances. My personal observation of laboratory, greenhouse, and growth chamber experiments is that large error variance is usually present, traits measured are of unknown or at best shaky correlation with field performance, and it takes a long time to conduct those experiments. To conduct indirect selection on more than one or two economic problems would strain a breeding program. Such an effort would become more complex when (and if) specific germplasm was documented for several component goals (e.g., salt, heat, cold, and shade tolerance) expected of a new turfgrass. Then, and only then, the turfgrass breeder would be left with the chore of recombining all these traits in one germplasm, preserving several traits among the recombinants, but not losing generalized adaptive traits or turfgrass quality.

This does not mean to say that we should not look for adaptive variants. Rather, such a search would be most productive when it uses the lawn or the field plot as the basis for defining adaptive criteria, not the other way around. An approach to shade tolerance, for example, would be to employ a naturally occurring shade environment while selecting for turfgrass quality. If factors other than light level (e.g., light quality, leaf litter, lepidopteran larvae, and root competition) were important, then their effects would be properly weighted by the turfgrass response, and its general performance would be what we would measure. Our goal is adaptation to the lawn or the golf course, not to the growth chamber. In addition, the breeder must carefully select the goals for indirect selection.

Selection Goals

Realistic goals are needed for the 21st century. Most effective crop improvement has been in the tailoring or the alteration of adapted plant materials to satisfy society's particular needs, rather than in the complete reshaping of a plant's physiology. In warm-season turfgrasses, limited resources have been available for breeding, and it is recommended that they be targeted in a reasonable manner.

Much effort in all warm-season turfgrasses has been directed towards the development of cold resistance. This is puzzling to understand, because, at the same time, cool-season grass breeders have been concerned with the development of heat tolerance. Some notable examples of cold resistance exist (e.g., Raleigh and Texas Common St. Augustinegrasses), but taxonomic studies show that what has been done, effectively, is the selection of a taxonomic group (Gulf Coast Group) which is morphologically distinctive from other groups. On the other hand, for species that are probably already being grown outside their natural range of adaptation, are there not other species that might be more serviceable? The conclusions here are that a little taxonomic knowledge can go far towards refining the worthwhile germplasm for selection, and conversely that some goals are not appropriate for all germplasms.

Shade tolerance will remain critical in warm-season turfgrasses. This must not be sidestepped by a general recommendation for groundcover or wood chips underneath trees. People will always want to enjoy themselves outdoors, on the grass, and underneath trees. The breakdown in the Earth's ozone layer, with a resulting influence in increasing skin cancer, makes it imperative to develop shade-tolerant turf cultivars. Shade tolerance is a goal for which natural variation has been observed, but has not been adequately developed, and major efforts are necessary. A better characterization of typical tree shade and of shade tolerance would be a good area for research.

Drought resistance of warm-season turfgrass will be important in the immediate future. Taxonomically oriented selection may help refine populations or races of merit for this trait, and may partially obviate the need for specific screening. There are already some races and species that can be grown without supplemental irrigation, and which provide nearly acceptable turf. Would it not be appropriate to try to select for better quality in already drought-resistant plant material, than to take species outside of their ranges of adaptation and to try to accomplish what millennia of natural selection have failed to do?

The conception of turfgrass drought resistance might be different, in the semiarid southwestern USA, where water is scarce, than in the humid southeast, where water is plentiful but seasonal. In the former, the breeder would look towards reduced evapotranspiration, since any deficit would eventually have to be made up for by irrigation. But in the latter, survival in the absence of irrigation might be feasible, because a genotype with a sufficiently deep root system might be able to reach a greater storage reservoir, and thereby withstand periodic droughts. The concept of drought resistance will direct the method of search, and whether or not the result is at all appropriate for the practical goal.

The search for resistance to acute pest damage must be emphasized, because this is a feasible and worthwhile goal. Chemical controls are limited in their applicability, both in life-span and in environmental safety. Genetic resistance offers the alternative of greater continuity, where pesticides have been banned or pests have become resistant. While we cheer the successes of host plant resistance, for example, St. Augustinegrass resistance to the southern chinch bug, the pest has become adapted to the resistant grass (Busey & Center, 1987). This must not dissuade us from the goal. Rather, it asks us to reaffirm a commitment to a safe and healthy environment that adapted turfgrasses can provide, and to science by which we will create them.

REFERENCES

Ahn, B.J., F.H. Huang, and J.W. King. 1985. Plant regeneration through somatic embryogenesis in common bermudagrass tissue culture. Crop Sci. 25:1107–1109.

Allen, R.J., Jr., and R.W. Kidder. 1971. Origin and history of Roselawn St. Augustinegrass. Proc. Soil Crop Sci. Soc. Fla. 30:354–360.

Atilano, R.A., and P. Busey. 1983. Susceptibility of St. Augustinegrass germ plasm to *Pyricularia grisea*. Plant Dis. 67:782–783.

Anonymous. 1903–1974. Plant inventory. USDA, Washington, DC.

Anonymous. 1976. Florida turfgrass survey 1974. Florida Dep. of Agric. and Consumer Serv., Tallahassee.

Barrios, E.P., F.J. Sundstrom, D. Babcock, and L. Leger. 1986. Quality and yield response of four warm-season lawngrasses to shade conditions. Agron. J. 78:270–273.

Biran, I., B. Bravdo, I. Bushkin-Harav, and E. Rawitz. 1981. Water consumption and growth rate of 11 turfgrasses as affected by mowing height, irrigation frequency, and soil moisture. Agron. J. 73:85–90.

Bogdan, A.V. 1970. Turfgrasses in Kenya. p. 51–56. *In* Proc. 1st Int. Turfgrass Res. Conf., Harrogate, England. 15–18 July 1969. Sports Turf Res. Inst., Bingley, England.

Bouton, J.H., A.E. Dudeck, R.L. Smith, and R.L. Green. 1983. Plant breeding characteristics relating to improvement of centipedegrass. Proc. Soil Crop Sci. Soc. Fla. 42:53–58.

Boutton, T.W., G.N. Cameron, and B.N. Smith. 1978. Insect herbivory on C_3 and C_4 grasses. Oecologia 36:21–32.

Bruton, B.D., R.W. Toler, and J.A. Reinert. 1983. Combined resistance in St. Augustinegrass to the southern chinch bug and the St. Augustine decline strain of panicum mosaic virus. Plant Dis. 67:171–172.

Burton, G.W. 1946. Bahia grass types. Agron. J. 38:273–281.

Burton, G.W. 1948. The method of reproduction in common bahia grass, *Paspalum notatum*. Agron. J. 40:443–452.

Burton, G.W. 1951. The adaptability and breeding of suitable grasses for the southeastern states. Adv. Agron. 3:197–241.

Burton, G.W. 1955. Breeding Pensacola bahiagrass, *Paspalum notatum*: I. Method of reproduction. Agron. J. 47:311–314.

Burton, G.W. 1967. A search for the origin of Pensacola bahia grass. Econ. Bot. 21:379–382.

Burton, G.W. 1974. Breeding bermudagrass for turf. p. 18–22. *In* E.C. Roberts (ed.) Proc. 2nd Int. Turfgrass Res. Conf., Blacksburg, VA. 19–21 June 1972. ASA and CSSA, Madison, WI.

Burton, G.W. 1982. Improved recurrent restricted phenotypic selection increases bahiagrass forage yields. Crop Sci. 22:1058–1061.

Burton, G.W. 1985. Registration of Tifway II bermudagrass. Crop Sci. 25:364.

Burton, G.W., E.H. DeVane, and R.L. Carter. 1954. Root penetration, distribution and activity in southern grasses measured by yields, drought symptoms and P^{32} uptake. Agron. J. 46:229–233.

Burton, G.W., and I. Forbes, Jr. 1960. The genetics and manipulation of obligate apomixis in common bahia grass (*Paspalum notatum* Flugge). p. 66–71. *In* Proc. 8th Int. Grass. Congr. Alden Press, Oxford, England.

Busey, P. 1980. Gamma ray dosage and mutation breeding in St. Augustinegrass. Crop Sci. 20:181–184.

Busey, P. 1985. Selection for seedhead characteristics in bahiagrass. p. 114. *In* Agronomy abstracts. ASA, Madison, WI.

Busey, P. 1986a. Bermudagrass germplasm adaptation to natural pest infestation and suboptimal nitrogen fertilization. J. Am. Soc. Hortic. Sci. 111:630–634.

Busey, P. 1986b. Morphological identification of St. Augustinegrass cultivars. Crop Sci. 26:28–32.

Busey, P., T.K. Broschat, and B.J. Center. 1982a. Classification of St. Augustinegrass. Crop Sci. 22:469–473.

Busey, P., and B.J. Center. 1987. Southern chinch bug (Hemiptera: Heteroptera: Lygaeidae) overcomes resistance in St. Augustinegrass. J. Econ. Entomol. 80:608–611.

Busey, P., and B.J. Myers. 1979. Growth rates of turfgrasses propagated vegetatively. Agron. J. 71:817–821.

Busey, P., J.A. Reinert, and R.A. Atilano. 1982b. Genetic and environmental determinants of zoysiagrass adaptation in a subtropical region. J. Am. Soc. Hortic. Sci. 107:79–82.

Carrier, L. 1927. Florida greenkeeping. Natl. Greenkeeper (Oct.) 1927:20, 25.

Chase, A. 1938. The carpet grasses. J. Wash. Acad. Sci. 28:178–182.

Dudeck, A.E. 1974. Flowering response of several selections of St. Augustinegrass. p. 74–78. *In* E.C. Roberts (ed.) Proc. 2nd Int. Turfgrass Res. Conf., Blacksburg, VA. 19–21 June 1973. ASA and CSSA, Madison, WI.

Dudeck, A.E., and C.H. Peacock. 1985. Effects of salinity on seashore paspalum turfgrasses. Agron. J. 77:47–50.

Dudeck, A.E., S. Singh, C.E. Giordano, T.A. Nell, and D.B. McConnell. 1983. Effects of sodium chloride on *Cynodon* turfgrasses. Agron. J. 75:927–930.

Dunn, J.H., K.M. Scheffer, and P.M. Halisky. 1981. Thatch and quality of Meyer zoysia in relation to management. Agron. J. 73:949–952.

Enlow, C.R., and W.E. Stokes. 1929. Lawns in Florida. Florida Agric. Exp. Stn. Bull. 209.

Fairchild, D. 1920. Plant inventory. (Comments on Plant Introductions 42839 and 42678, zoysiagrass; and 46415, centipedegrass). USDA Agric. Res. Serv. (Formerly, Inventory, Bureau of Plant Industry).

Fairchild, D. 1938. The world was my garden: Travels of a plant explorer. Charles Scribners Sons, New York.

Finlayson, E.H. 1941. Pensacola—a new fine leafed bahia. South. Seedsman (Dec) 1941:9, 28.

Flint, C.L. 1903. Grasses and forage plants. Lee and Shepard, Boston.

Forbes, I. 1952. Chromosome numbers and hybrids in *zoysia*. Agron. J. 44:194–199.

Forbes, I. 1962. Registration of Emerald zoysiagrass. Crop Sci. 2:533–534.

Forbes, I., Jr., and G.W. Burton. 1963. Chromosome numbers and meiosis in some *Cynodon* species and hybrids. Crop Sci. 3:75–79.

Fosberg, F.R. 1977. Paspalum distichum again. Taxon 26:201–202.

Goudswaard, P.C. 1980. The genus *Zoysia* (Gramineae) in Malesia. Blumea 26:169–175.

Gould, F.W. 1968. Grass systematics. McGraw-Hill, New York.

Gray, L.C. 1958. History of agriculture in the southern United States to 1860. Peter Smith, Gloucester, Mass. Reprinted as Publ. 430. Carnegie Inst. of Washington, Washington, DC.

Greenfield, I. 1974. Turfgrass developments in southeast Asia. p. 438–444. *In* E.C. Roberts (ed.) Proc. 2nd Int. Turfgrass Res. Conf., Blacksburg, VA. 19–21 June 1973. ASA and CSSA, Madison, WI.

Grisham, M.P., J.L. Dale, and R.D. Riggs. 1974. *Meloidogyne graminis* and *Meloidogyne* spp. on *Zoysia*: Infection, reproduction, disease development, and control. Phytopathology 64:1485–1489.

Hanna, W.W., and G.W. Burton. 1978. Cytology, reproductive behavior, and fertility characteristics of centipedegrass. Crop Sci. 18:835–837.

Hanna, W.W., and H. Shu-Liang. 1986. Chromosomal, reproductive, and breeding behavior of *Eremochloa ciliaris* (L.) Merr. J. Hered. 77:52–53.

Hanson, A.A. 1972a. Breeding of grasses. p. 36–52. *In* V.B. Youngner and C.M. McKell (ed.) The biology and utilization of grasses. Academic Press, New York.

Hanson, A.A. 1972b. Grass varieties in the United States. USDA Agric. handb. 170. U.S. Gov. Print. Office, Washington, DC.

Hanson, A.A., F.V. Juska, and G.W. Burton. 1974. Species and varieties. *In* A.A. Hanson and F.V. Juska (ed.) Turfgrass science. Agronomy 14:370–409.

Harlan, J.R. 1975. Crops and man. ASA and CSSA, Madison, WI.

Harlan, J.R., and J.M.J. de Wet. 1969. Sources of variation in *Cynodon dactylon* (L.) Pers. Crop Sci. 9:774–778.

Harlan, J.R., J.M.J. de Wet, W.W. Huffine, and J.R. Deakin. 1970a. A guide to the species of *Cynodon* (Gramineae). Oklahoma Agric. Exp. Stn. Bull. B-673.

Harlan, J.R., J.M.J. de Wet, K.M. Rawal, M.R. Felder, and W.L. Richardson. 1970b. Cytogenetic studies in *Cynodon* L.C. Rich. (Gramineae). Crop Sci. 10:288–291.

Horn, G.C., A.E. Dudeck, and R.W. Toler. 1973. 'Floratam' St. Augustinegrass: A fast growing new variety for ornamental turf resistant to St. Augustine decline and chinch bugs. Florida Agric. Exp. Stn. Circ. S-224.

Huff, D.R., and L. Wu. 1987. Sex expression in buffalograss under different environments. Crop Sci. 27:623–626.

Ibitayo, O.O., J.D. Butler, and M.J. Burke. 1981. Cold hardiness of bermudagrass and *Paspalum vaginatum* Sw. HortScience 16:683–684.

Jean, S., and T. Juang. 1979. Effect of bahia grass mulching and covering on soil physical properties and losses of water and soil of slopeland (First report). J. Agric. Assoc. China 105:57–66.

Johnson, W.J., and R. Dickens. 1977. Cold tolerance evaluation of several centipedegrass selections. Agron. J. 69:100–103.

Johnson, W.J., R. Dickens, R. Haaland, and J.R. Cooper. 1977. Irradiation-induced variation in centipedegrass. p. 60 *In* Agronomy abstracts. ASA, Madison, WI.

Juska, F.V., and J.J. Murray. 1974. Performance of bermudagrasses in the transition zone as affected by potassium and nitrogen. p. 149–154. *In* E.C. Roberts (ed.) Proc. 2nd Int. Turfgrass Res. Conf., Blacksburg, VA. 19–21 June 1973. ASA and CSSA, Madison, WI.

Kakuda, S. 1987. Overview of the turfgrass industry in Japan. Harvests 34:3. The Lawn Inst., Pleasant Hill, TN.

Killinger, G.B., G.E. Ritchey, C.B. Blickensderfer, and W. Jackson. 1951. Argentine bahia grass. Florida Agric. Exp. Stn. Circ. S-31.

Kitamura, F. 1970. Studies on the horticultural classification and development of Japanese lawn grasses. Bull. Kemigawa Arboretum, Fac. of Agric. Univ. of Tokyo 3:1–60.

Krans, J.V., J.B. Beard, and J.F. Wilkinson. 1979. Classification of C_3 and C_4 turfgrass species based on CO_2 compensation concentration and leaf anatomy. HortScience 14:183–185.

Latham, J.M., Jr. 1966. Better bermudagrass greens and tees. Milwaukee Sewerage Commission Bull. 6.

Leuck, D.B., C.M. Taliaferro, G.W. Burton, R.L. Burton, and M.C. Bowman. 1968. Resistance in bermudagrass to the fall armyworm. J. Econ. Entomol. 61:1321–1322.

Maki, Y. 1976. Utilization, research activities and problems of turfgrass in Japan. Rasen-Turf-Gazon 7:24–28.

Martin, R.J. 1975. A review of carpet grass (*Axonopus affinis*) in relation to the improvement of carpet grass based pasture. Trop. Grassl. 9:9–19.

McBee, G.G., and E.C. Holt. 1966. Shade tolerance studies on bermudagrass and other turfgrasses. Agron. J. 58:523–525.

Mills, P.F.L., and J.N. Boultwood. 1978. A comparison of *Paspalum notatum* accessions for yield and palatability. Zimbabwe Agric. J. 75:71–74.

Moraes, Fernandes, M.I.B., I.L. Barreto, and F.M. Salzano. 1973. Cytogenetic, ecologic and morphologic studies in Brazilian forms of *Paspalum notatum*. Can. J. Genet. Cytol. 15:523–531.

Morton, J.F. 1973. Salt-tolerant silt grass (*Paspalum vaginatum* Sw.). Proc. Fla. State Hortic. Soc. 86:482–490.

Nutter, G.C., and R.J. Allen, Jr. 1960. Floratine St. Augustinegrass: A new variety for ornamental turf. Florida Agric. Exp. Stn. Circ. S-123.

Ohwi, J. 1965. Flora of Japan. Smithsonian Inst., Washington, DC.

Palmertree, H.D., C.Y. Ward, and R.H. Pluenneke. 1974. Influence of mineral nutrition on the cold tolerance and soluble protein fraction of centipedegrass. p. 500–507. *In* E.C. Roberts (ed.) Proc. 2nd Int. Turfgrass Res. Conf., Blacksburg, VA. 19–21 June 1973. ASA and CSSA, Madison, WI.

Parodi, L.R. 1948. Gramineas Argentinas nuevas o críticas: 1. La variación en *Paspalum notatum* Fluegge. Rev. Argent. Agron. 15:53–61.

Peacock, C.H., and A.E. Dudeck. 1981. The effects of shade on morphological and physiological parameters of St. Augustinegrass cultivars. p. 493–500. *In* R.W. Sheard (ed.) Proc. 4th Int. Turfgrass Res. Conf., Guelph, Ontario, Canada. 19–23 July. Ontario Agric. Coll., Univ. Guelph and Int. Turfgrass Soc., Guelph, Ontario.

Pedersen, J.F., and R. Dickens. 1985. Registration of AU Centennial centipedegrass. Crop Sci. 25:364.

Piper, C.V. 1915. P.I. 34657. *Osterdamia matrella* (L.) Kuntze (*Zoysia pungens* Willd.). *In* Plant inventory 33:43. USDA, Washington, DC.

Powell, J.B., G.W. Burton, and J.R. Young. 1974. Mutations induced in vegetatively propagated turf bermudagrasses by gamma radiation. Crop Sci. 14:327–330.

Powell, J.B. and R.W. Toler. 1980. Induced mutants in 'Floratam' St. Augustinegrass. Crop Sci. 20:644–646.

Quarin, C.L., B.L. Burson, and G.W. Burton. 1984. Cytology of intra- and interspecific hybrids between two cytotypes of *Paspalum notatum* and *P. cromyorrhizon*. Bot. Gaz. 145:420–426.

Reeves, S.A., Jr., and G.G. McBee. 1972. Nutritional influences on cold hardiness of St. Augustinegrass (*Stenotaphrum secundatum*). Agron. J. 64:447–450.

Reinert, J.A., B.D. Bruton, and R.W. Toler. 1980. Resistance of St. Augustinegrass to southern chinch bug and St. Augustine Decline Strain of Panicum Mosaic Virus. J. Econ. Entomol. 73:602–604.

Reinert, J.A., and P. Busey. 1983. Resistance of bermudagrass selections to the tropical sod webworm (Lepidoptera: Pyralidae). Environ. Entomol. 12:1844–1845.

Reinert, J.A., and P. Busey. 1984. Resistant varieties. p. 35–40. *In* T.J. Walker (ed.) Mole crickets in Florida. Florida Agric. Exp. Stn. Bull. 846.

Reinert, J.A., P. Busey, and F.G. Bilz. 1986. Old world St. Augustinegrasses resistant to the southern chinch bug (Heteroptera: Lygaeidae). J. Econ. Entomol. 79:1073–1075.

Reinert, J.A., and A.E. Dudeck. 1974. Southern chinch bug resistance in St. Augustinegrass. J. Econ. Entomol. 67:275–277.

Reinert, J.A., A.E. Dudeck, and G.H. Snyder. 1978. Resistance in bermudagrass to the bermudagrass mite. Environ. Entomol. 7:885–888.

Reinert, J.A., R.W. Toler, B.D. Bruton, and P. Busey. 1981. Retention of resistance by mutants of 'Floratam' St. Augustinegrass to the southern chinch bug and St. Augustine Decline. Crop Sci. 21:464–466.

Riodan, T.P., V.D. Meier, J.A. Long, and J.T. Gruis. 1980. Registration of Seville St. Augustinegrass. Crop Sci. 20:824–825.

Sakai, K. 1983. New summer crop cultivars (I)—New cultivars registered in Ministry of Agriculture, Forestry and Fisheries in 1983 fiscal year. Jpn. J. Breed. 33:499–506.

Sauer, J.D. 1972. Revision of *Stenotaphrum* (Gramineae: Paniceae) with attention to its historical geography. Brittonia 24:202–222.

Simmonds, N.W. 1972. Principles of crop improvement. Longman, New York.

Tarjan, A.C., and P. Busey. 1985. Genotypic variability in bermudagrass damage by ectoparasitic nematodes. HortScience 20:675–676.

Thompson, W.R., Jr. 1965. Vegetative identification of 16 varieties of turf-type bermudagrass (*Cynodon* spp.). Bull. Fla. Turf-Grass Assoc. 12(1):6–8.

Tracy, S.M. 1917. Bermuda grass. USDA Farmers' Bull. 814. USDA, Washington, DC.

Voigt, P.W., W.R. Kneebone, J.R. Harlan, and R.M. Ahring. 1975. Registration of Texoka buffalograss. Crop Sci. 15:885.

Walker, J.T., and R.E. Burns. 1975. Centipedegrass decline in Georgia. Georgia Agric. Exp. Stn. Res. Bull. 167.

White, R.W., and P. Busey. 1987. History of turfgrass production in Florida. Proc. Fla. State Hortic. Soc. 100:167–174.

Wilson, C.A., J.A. Reinert, and A.E. Dudeck. 1977. Winter survival of St. Augustine grasses in North Mississippi. S. Nurserymen's Assoc. Res. Conf. Annu. Rep. 22:195–198.

Winstead, C.W., and C.Y. Ward. 1974. Persistence of southern turfgrasses in a shade environment. p. 221–230. *In* E.C. Roberts (ed.) Proc. 2nd Int. Turfgrass Res. Conf., Blacksburg, VA. 19–21 June 1973.. ASA and CSSA, Madison, WI.

Wofford, D.S., and A.A. Baltensperger. 1985. Heritability estimates for turfgrass characteristics in bermudagrass. Crop Sci. 25:133–136.

Youngner, V.B. 1961. Growth and flowering of zoysia species in response to temperatures, photoperiods, and light intensities. Crop Sci. 1:91–93.

Youngner, V.B., and F.J. Nudge. 1980. Air pollution oxidant effects on cool-season and warm-season turfgrasses. Agron. J. 72:169–170.

6 Breeding Forage Grasses to Maximize Animal Performance[1]

Gordon C. Marten
USDA-ARS
University of Minnesota
St. Paul, Minnesota

ABSTRACT

Animal performance may depend heavily upon the true feeding value of a forage grass that can be expressed as the nutrients used per unit of time. Breeding of forage grasses to improve animal performance has been documented to be successful in several cases where breeders have selected germplasm for greater leafiness, improved digestibility or palatability per se, and reduced alkaloids. Grass breeders must be alert to the influences of maturation, leafiness, stand condition, environment or environmental adaptation, capacity to mix with high-quality legumes, pest susceptibility, and potential antiquality constituents when they select for feeding value improvement. Yield need not be sacrificed in order to gain quality improvements in many instances, but small sacrifices in yield can be acceptable if quality improvements increase animal performance. Small improvements in digestibility or intake of a grass can result in large differences in animal performance. The influence of antiquality constituents can supersede the influence of digestibility and intake in some grass species. Excellent laboratory methods are available to aid breeders in screening grass germplasm for improved digestibility, reduced fiber-component concentration, and possibly improved intake potential. Near infrared reflectance spectroscopy (NIRS) has great potential as a fast and precise method for screening large numbers of genotypes for most feeding value traits that are quantifiable by current or future chemical, physical, or biological methods. New developments in the areas of cell wall carbohydrates, phenolic compounds, plant anatomy, protein degradability, mastication and particle flow, and other mechanical stability merit careful attention. We need a "bidirectional" selection approach, imaginative selection criteria based on theoretical as well as proven concepts, and an interdisciplinary research effort to hasten progress in breeding improved grasses to maximize animal performance.

"Grass bears no blazonry of bloom—it yields no fruit in earth or air, yet should its harvest fail for a single year, famine would depopulate the world."[2] This statement by Senator Ingalls in 1885 recognized the crucial role that forage grasses play worldwide in sustaining our food supply via ruminant animal products as well as in sustaining our natural resources.

[1] Contribution of the USDA-ARS in cooperation with the Minnesota Agric. Exp. Stn. Scientific Journal Series Paper no. 15 856.
[2] Statement by Senator Ingalls of Kansas in 1885 (from CAST, 1986).

 Contributions from Breeding Forage and Turf Grasses, CSSA Special Publication no. 15.

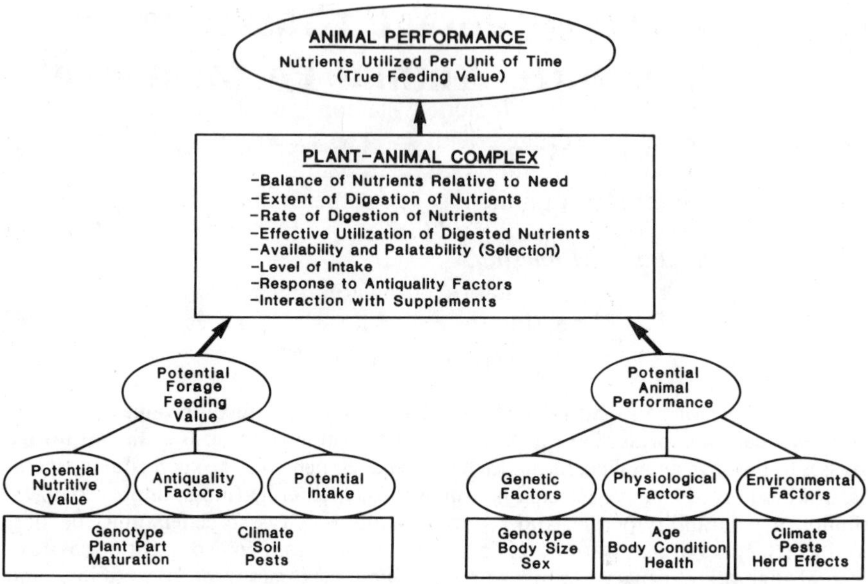

Fig. 6-1. Plant and animal factors that influence feeding value (quality) of forages.

DEFINITION OF FORAGE QUALITY (FORAGE FEEDING VALUE)

The complex of plant and animal factors that contributes to performance of ruminants that consume forage grasses as part or all of their energy, protein, and mineral nutrients is presented in Fig. 6-1 (Marten, 1985a). True feeding value of a forage is determined by intake, digestibility, and effective utilization of specific nutrients in relation to the specific needs of the animal; potential feeding value may also be modified by inherent antiquality factors that are either toxic or that interfere with intake, digestibility, or effective utilization of nutrients. Whereas relatively small differences in nutritive value and/or intake of grasses can cause large differences in animal performance, the impact of antiquality constituents can be so great that conventional measurement of potential quality becomes irrelevant.

The most critical genetic traits that influence the animal performance potential of forage grasses are their proportion of plant parts (morphology), maturation rate, inherent chemical and anatomical composition (nutritive value and intake potential per se), pest resistance, capacity to coexist in mixture with high-quality legumes, and inherent antiquality traits. Examples of how selection to regulate these genetic traits in forage grasses may influence ruminant animal performance, or how they have been documented to influence performance, will be considered.

PREVIOUS REVIEWS

In 1976, Div. C-1, C-2, C-3, and C-6 of the Crop Science Society of America conducted a symposium, "Breeding for Forage Quality", that included unpublished papers entitled, "Breeding to Increase Quality in Temperate Grasses" (Asay, 1976) and "Breeding to Improve Quality in Tropical Grasses" (Burton, 1976). Asay concluded that a major goal was the development of a selection index that included the combination of plant traits associated with quality, forage yield, pest resistance, and tolerance to environmental stress that would facilitate maximum expression of available genetic variation in terms of animal products per unit area of land. He also presented breeding schemes that would take maximum advantage of new innovations such as NIRS.

Burton (1976) concluded that breeding could improve forage quality in tropical grasses by (i) increasing their resistance to diseases and insects; (ii) "restructuring" them morphologically; (iii) increasing their intake, digestibility, and nutrient concentration; and (iv) altering their photoperiodic response.

Other major reviews of progress and potentials in breeding of forage grasses for improved feeding value to ruminant animals include those of Shenk (1977), van Bogaert (1977), Burton (1974, 1986), and Hacker (1982). The review by Hacker (1982) is especially comprehensive in that he delineated the evidence up until about 1979 of considerable genetic variation within numerous grass species for digestibility; concentration of soluble carbohydrates, fiber components, minerals, and toxins; and intake potential. He presented heritability estimates and phenotypic and genotypic relationships between feeding value characteristics that revealed that plant breeders should be able to improve the quality of forage grasses. Soh et al. (1984) also reviewed recent literature concerning genetic variation within several grass species for fiber components, digestibility, and protein.

SPECIAL CONCERNS FOR GRASS BREEDERS

Shenk (1977) pointed out that three major areas of concern for future grass breeding programs directed toward forage quality improvement were: (i) genotype × environmental interactions, (ii) spaced plant selection vs. solid-seeded stand evaluation, and (iii) the influence of maturity on quality among genotypes. He concluded that a possible selection scheme, in view of these concerns, would be to progeny test large populations of parental clones over locations, harvests, and generations in solid-seeded mini-plots.

Vogel et al. (1986), however, concluded that although significant genotype × environment interactions occurred for yield of intermediate wheatgrass [*Thinopyrum intermedia* (Host) Barkworth and D.R. Dewey], its concentration of in vitro digestible dry matter (IVDDM) was relatively stable over environments. Also, Burner et al. (1983) reported that interactions between tall fescue (*Festuca arundinacea* Schreb.) genotypes and environments for water-soluble carbohydrate concentration were minor.

Ugherughe et al. (1980) concluded that for an evaluation method used in grass breeding to be of much practical value, it had to correctly appraise the performance of the forage under the system of management that would be used in farming systems. They found that smooth bromegrass (*Bromus inermis* Leyss.) digestibility estimates based on samples obtained from spaced plants differentiated among genotypes as well as, or better than, digestibility estimates based on samples taken from solid stands. Because spaced plants offered a "better working environment to the forage breeder," their use in estimating digestibility was considered valid provided the selections ultimately produced superior animal performance.

EXAMPLES OF HOW SELECTION FOR TRAITS IN FORAGE GRASSES MAY POTENTIALLY INFLUENCE RUMINANT ANIMAL PERFORMANCE

Morphology and Maturation

Leafiness, Plant Part Digestibility, and Intake

In a comprehensive review of the digestibility of plant parts, Hacker and Minson (1981) revealed that *among* grass species, the mean digestibility of leaf blades (lamina) ranged from 46 to 81% of the dry matter, with leaves of tropical species tending to be the least digestible. No clear evidence existed, however, for interspecific differences in digestibility of grass leaf sheaths or stems. *Within* grass species, the mean digestibility varied 28% units for leaf sheaths, 22% units for stems, and 18% units for leaf blades. At a comparable stage of development, grass leaf sheaths were often 10% units less digestible than leaf blades. Generally, grass stems have a lower feeding quality than leaves or reproductive organs, except at early stages of maturation. Therefore, leafiness of perennial grasses is usually highly positively associated with digestibility of the total forage (Hacker & Minson, 1981; Christensen et al., 1984).

Minson (1983) pointed out that in most tropical pastures and in mature temperate pastures, large-quality differences exist between leaf and stem fractions. Stem feeding value is low and greater animal energy is required to sever the stem compared to the leaf. Also, grazing ruminants search for leaves even when few are present, which may cause low intake. Further, Minson (1980) reported that for five tropical grasses, physical measurements for leaf percentage, surface area, density, or grinding energy were better predictors (in conjunction with digestibility) of intake than was chemical composition. Leaf percentage was the best of all additional predictors of intake ($r = 0.87$). In addition, Minson (1982) cited evidence that leaf fractions were eaten by ruminants in greater quantities than were stem fractions of similar digestibility. The higher intake of the leaf fraction was associated with shorter rumen retention time. The most probable reason for the longer rumen retention time of the stem fractions was the greater proportion of large particles in masticated stem than in masticated leaf, because of the greater resistance of stem tissue to physical breakdown.

As a result, genetic selection for increased leaf blade fraction, reduced stem fraction, or for increased stem and leaf digestibility (individually) has become a plant breeding goal. In 1962, Sullivan et al. investigated the composition and in vitro digestibility of nonheading orchardgrass (*Dactylis glomerata* L.) germplasm. They reported that, in spring growth, chemical changes in nonheading plants proceeded in the same direction as normal plants, but the changes of nonheading plants were less extensive. By head emergence of the normal plants, the nonheaders had about 3.5% units more crude protein, and by bloom of the normal plants, the nonheaders had more than 1% unit less lignin and were almost 18% units more digestible. The nonheading plants, however, usually yielded less DM and could not be propagated except by vegetative cuttings. From the nonheading germplasm, Hovin et al. (1966) produced an essentially nonflowering orchardgrass synthetic that had improved forage quality, but which yielded only about 65 to 80% as much forage as a commercial cultivar. In follow-up research, Berg et al. (1981) found that sparse-flowering orchardgrass synthetics had slightly lower first harvest forage yields than control cultivars, but no differences in IVDDM or lignin concentration.

Marum and Hovin (1979) found that leafy plants of reed canarygrass (*Phalaris arundinacea* L.) had less total herbage cell wall constituent (CWC) concentration and often less CWC in all plant parts than did stemmy plants. Path-coefficient analysis revealed that CWC in the stem had most influence on total herbage CWC at anthesis, but CWC in the leaf blade had most influence on CWC in vegetative regrowth. Concentration of CWC in stems and blades was highly positively correlated, and DM yield was positively correlated with total herbage CWC concentration. In subsequent studies with reed canarygrass, Christensen et al. (1984) found that simultaneous selection for increased leaf blade fraction, visual leaf rigidity, and tiller number may be an effective way to identify high-yielding genotypes that are also highly digestible. On the other hand, Greub et al. (1986) reported that whereas sufficient genetic variation existed within quackgrass [*Elytrigia repens* (L.) Nevski] to enable selection to improve IVDDM, crude protein, and perhaps neutral detergent fiber (NDF), leafiness, growth habit, and other morphological traits were not sufficiently well correlated with forage quality to enable their use in selection to improve quality. Hacker (1982) cited several other exceptional instances where leafiness of grass species was not well associated with their digestibility or intake potential.

A striking example of how breeding to alter leafiness and plant part proportions may influence grass feeding value and ruminant animal performance is provided by the research of Burton et al. (1969). When they reduced the height of the pearl millet [*Pennisetum americanum* (L.) K. Schum.] cv., Tifleaf 1, with the d_2 gene, they reduced yields by 22%, but increased leafiness by 50%. This resulted in 21% increase in the intake of dehydrated grass and 49% increase in daily gains by heifers (*Bos taurus*); it also allowed grazing steers to gain 20% more compared to those that grazed a similar tall hybrid when equivalent stocking rates were used (Table 6–1).

Generally, there is less variation in grass leaf digestibility than in stem digestibility, and heritability estimates for leaf digestibility are lower (Hacker, 1982). But, leaf and stem digestibility are usually positively correlated

Table 6-1. Performance of near-isogenic tall and dwarf pearl millet fed to cattle. Adapted from Burton et al. (1969).

Measurement	Tall	Dwarf	Dwarf as a percentage of tall
	Dehydrated, boot-stage grass		
Dry matter yield, kg/ha	7200	5600	78
Dry matter intake, kg/heifer per day	5.19	6.29	121
Average daily gain, kg/heifer	0.35	0.52	149
kg dry matter/kg gain	15	12	80
Heifer gain, kg/ha	480	467	97
	Grazed grass		
Steer days/ha	654	556	85
Avg. daily gain, kg/steer	0.59	0.71	120
Steer gain, kg/ha	386	393	102

within grass genotypes. Carlson et al. (1983) reported that specific leaf weight (dry weight per unit leaf area) had no consistent influence on IVDDM, NDF, acid detergent fiber (ADF), hemicellulose, cellulose, lignin, silica, or insoluble ash concentrations in nonflowering regrowth of reed canarygrass. Whereas research on C_4, C_3, and intermediate photosynthetic types of the *Panicum* genus has revealed variations in leaf blade anatomy associated with digestibility, Akin et al. (1984) found no distinctive stem anatomical variations relative to rumen microbial degradation or lignification patterns that could be associated with photosynthetic type.

Removal of the waxy bloom from leaves of sorghum [*Sorghum bicolor* (L.) Moench] via single gene selection resulted in a 31% increase in IVDDM (Hanna et al., 1974). They also reported that suppression of shoot trichomes via a single gene mutation of normal (hairy) pearl millet caused it to have a more "waxy" surface, and that leaves of the normal type were 31% more digestible than those of the mutant. They suggested that in areas where water is not limiting to plant growth, the bloomless gene could allow improvements in animal performance through more rapid digestion and increased intake of the forage. Hacker (1982) suggested selection for low levels of cutinization of grass leaves may have practical benefits, depending on possible deleterious side-effects such as increased transpiration, reduced yield, and increased pest susceptibility.

Maturation

Mature grasses are usually less digestible, have less protein and minerals, more fiber and lignin, and lower intake potential than immature grasses. These changes partly reflect the increase in proportion of stem tissue that not only contains more lignin than the leaves, but undergoes greater lignification during maturation. With advancing maturity of grasses, lignification of specific cells within the vascular tissue is closely associated to the feeding value decline (Akin et al., 1977). The IVDDM of both leaves and stems declines sharply in most grasses as they mature and as fibrous components increase, with the extent of the decline nearly always being greater for the stems. Morrison (1980) found that a strong correlation ($r = 0.87$) existed between lignin and hemicellulose concentrations with increasing maturation of the herbage of six varieties of perennial ryegrass (*Lolium perenne* L.) and one variety each of timothy (*Phleum pratense* L.), orchardgrass, tall fescue, and meadow fescue (*Festuca pratensis* Hud.). In all cases, the lignin and hemicellulose increases were greater in the stems than in the leaves, and diploid cultivars of perennial ryegrass had higher lignin and hemicellulose concentrations than tetraploids.

When Holt et al. (1979) examined five clones from each of five kleingrass (*Panicum coloratum* L.) and seven buffelgrass (*Cenchrus ciliaris* L.) genotypes, they found substantial genotype × age interactions (from samples collected at three regrowth ages), which indicated the need for a breeder to sample at more than one stage of maturation to gain a superior assessment of IVDDM potential. Stratton et al. (1979) reported that whereas it should be possible to increase the IVDDM of first harvest orchardgrass herbage without sacrificing DM yield, it may be more difficult to increase IVDDM (or reduce fiber concentration) in regrowth herbage without a decrease in yield. Maturity was negatively associated with quality at the first harvest dates when all progeny in polycross blocks of 14 clones were harvested on the same six dates.

In experiments with 18 cultivars of annual ryegrass (*Lolium multiflorum* Lam.), Nelson and Rouquette (1983) found that differences in NDF concentration of the herbage were related to maturation in that earlier-maturing cultivars had higher NDF; tetraploids had lower NDF concentrations than diploids, probably because the tetraploids matured later. Similarly with perennial ryegrass, sheep (*Ovis aries*) preferred late cultivars to early ones, and tetraploids to diploids, in cafeteria trials; differences in growth stage of all cultivars had a much greater effect on sheep intake than did inherent genetic differences in cultivars (Simon & Daniel, 1983).

Holt and Conrad (1986) found that all cultivars of bermudagrass [*Cynodon dactylon* (L.) Pers.] did not respond the same to age and year effects on IVDDM, which suggested the possibility of altering these effects via breeding. One cultivar (S-83) tended to decrease in IVDDM less rapidly with advancing age. When Reich and Casler (1985) evaluated nine polycross families of a smooth bromegrass experimental synthetic for differences in cell wall fraction response to advancing maturity (RTM), they concluded that improved herbage quality would be expected from selection for a slow RTM for NDF, ADF, or cellulose. They also concluded that selection for a slower RTM would not cause a shift toward lateness in the populations, because the polycross families did not differ in maturity.

Burton (1976) found that short-day pearl millets matured much later, yielded more, provided a better seasonal distribution of forage, were leafier, more digestible, and had a greater crude protein concentration than their near-isogenic normal counterparts.

Certainly in breeding grasses for improved forage quality, the breeder must attempt to eliminate the effect of growth stage from selection comparisons. Maturity effects are dominant in initial spring growth; therefore, plant breeders will often find less problem with maturity variation in regrowth herbage, but even here carryover effects from first-crop differences may occur. If breeders select for improved feeding value at specific harvest dates without accounting for growth stage differences, they will likely select plants that are merely later maturing.

Nutritive Value and Intake Potential

Grasses vs. Legumes

Generally, forage grasses have lower feeding value than legumes harvested at comparable stages of maturation (Marten, 1985a). Whereas grasses and legumes frequently have similar dry matter digestibility at comparable stages, the grasses contain substantially more NDF (cell walls), their rate of DM digestion is slower and, therefore, their intake potential is less than that of legumes. Grasses and legumes have several major fiber composition differences that help explain the more rapid digestion and greater intake of legumes. At identical or similar digestibility, grasses usually contain considerably less lignin and considerably more hemicellulose in their cell walls than do legumes. Therefore, the cell wall fraction in legumes is decidedly less digestible than that in grasses. This lower digestibility of legume cell walls is offset by greater concentrations of nearly completely and quickly digestible cell solubles (less cell walls), so legumes and grasses often have the same digestibility at similar maturation stages. Unless grasses are grown with ample N fertilization, they will contain less crude protein (CP) than legumes, and they usually contain substantially less Ca and less total minerals than legumes.

Tropical vs. Temperate Grasses

Differences in the feeding value of tropical and temperate grasses identified by Mott (1983) were that temperate species generally have: (i) less crude fiber; (ii) less retention time in the reticulo-rumen; (iii) greater tiller density and bulk density in canopy segments during grazing (to facilitate larger bite sizes and greater intake); (iv) greater CP; and (v) greater digestibility. He cited data that showed that tropical grasses have voluntary intake rates by sheep that are 10 to 15 units (g/day per metabolic wt.) lower than those of temperate grasses. Minson (1982) also concluded that this intake difference appeared to be associated with the lower digestibility of tropical grasses, even though the relationship between intake and digestibility differs within tropical vs. temperate grasses as well.

Warm-season grasses had an average of 10% units more NDF and 4% units more ADF than cool-season grasses of similar maturation (Barton et al., 1976). However, most of the digestibility advantage for temperate grasses can apparently be attributed to their greater rate and extent of digestion of the cell wall. Tropical grasses have a more developed parenchyma bundle sheath compared to temperate grasses, and up to 25% of the cross-sectional area of leaf blades of warm-season grasses is occupied by this type of tissue (Akin & Burdick, 1975). The epidermis and parenchyma bundle sheath of tropical grasses are especially resistant to rumen microbial digestion. Thus, differences in leaf blade digestibility between temperate and tropical grasses are related to the relative ease of tissue digestion rather than to the amount of nondigestible, lignified tissue (Akin, 1983). Tissues in orchardgrass, a temperate grass, were rapidly degraded (including parenchyma bundle sheath and epidermis tissues). Those in bermudagrass, a tropical grass, had markedly slower digestion (Fig. 6–2). Also, more total bacteria adhered to the cell walls of orchardgrass during digestion in vitro. Histochemical studies revealed that tissues that were slowly or only partially degraded in bermudagrass were lig-

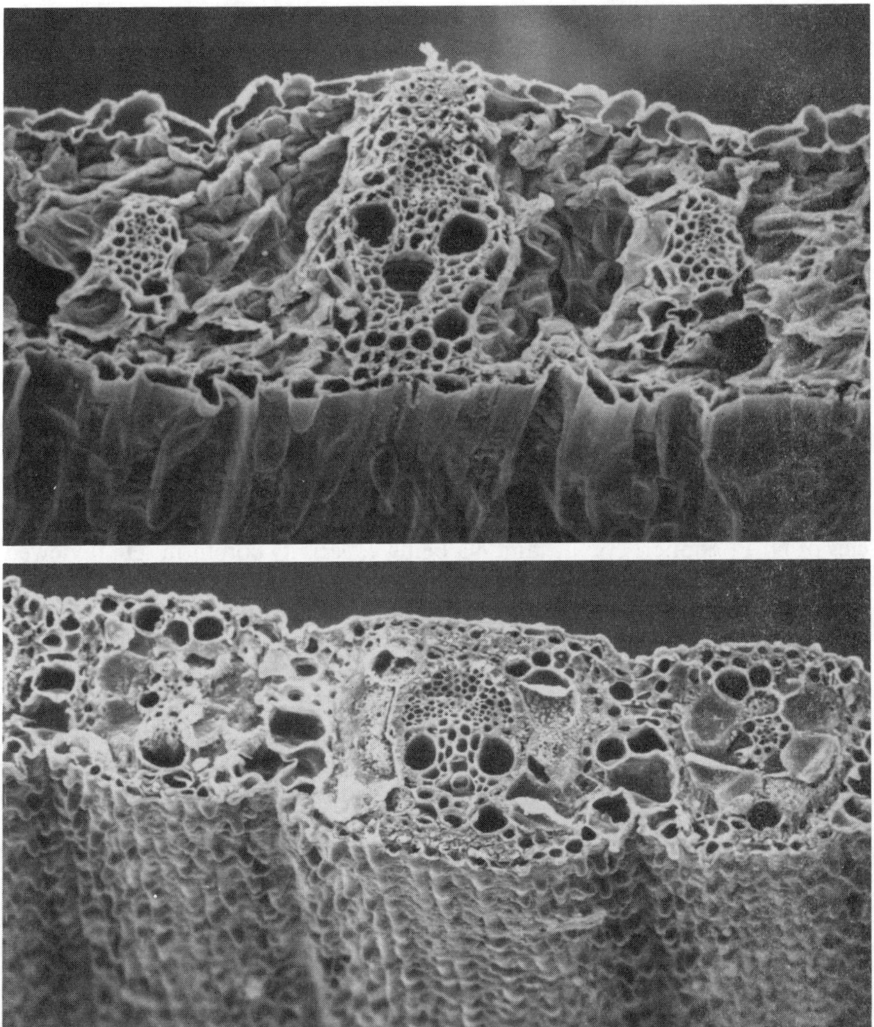

Fig. 6–2. Non-Kranz anatomy of cross-section of orchardgrass leaf blade (*top*) compared to Kranz anatomy of Coastal bermudagrass leaf blade (*bottom*). Courtesy of D.E. Akin, Athens, GA.

nified, which helped to explain the lack of availability of hemicellulose and other cell wall constituents to rumen microorganisms. Evidence cited by Morrison (1979) showed that, in grasses, part or all of the lignin is covalently bound to hemicellulose and that the acetyl groups are linked exclusively to hemicellulose.

The Nutritionally Ideal Grass

Minson (1981b) listed the following "desirable" and "undesirable" feeding value factors to use in the quest for a "nutritionally ideal grazing plant."

Desirable factors:

1. Mineral composition, to meet the needs of ruminants.
2. Protein, not lower than 7% and up to 16% dry wt. or more for high producers.
3. Energy digestibility, about 80%.

Undesirable factors:

1. Poor spatial distribution so as to be inaccessible (inadequate bite size).
2. Low intake due to excess indigestible or slowly digestible fiber (steminess).
3. Characters that reduce palatability.
4. Toxic and oestrogenic compounds.

Most of these attributes would also apply to a nutritionally ideal harvested forage grass.

Mineral Concentration and Balance

Hacker (1982) pointed out that low levels of essential minerals often limit the production of ruminants that consume tropical grasses, with deficiencies of N, P, copper, cobalt, and Na being especially common. He reviewed literature revealing that grasses usually have extensive genetic variation in concentration of major elements, and that heritability estimates for major elements are often high. He also found evidence of heritable variation for micronutrient concentration in grasses. He further reported frequent negative genetic correlations between yield and concentration of major minerals in grasses, and that selection to change concentration of an element is likely to cause changes in concentration of other elements. Gorz et al. (1987) recently showed the feasibility of altering the concentration of 12 minerals in forage sorghum by breeding. Nasinec and Mika (1980) recommended selection of orchardgrass parents for Na concentration before crossing.

Much of the recent research concerning the potential to alter mineral concentration of perennial temperate grasses has featured K, Ca, Mg, and especially the K/(Ca + Mg) ratio. Sleper et al. (1977) reviewed the evidence that a K/(Ca + Mg) ratio of 2.2 ($cmol_c\ kg^{-1}$) in grasses may increase the incidence of hypomagnesaemia (grass tetany) in ruminants that graze them. They and Reeder et al. (1986) reported that since the K/(Ca + Mg) ratio was highly heritable in tall fescue, it should be possible to breed cultivars that have low grass tetany potential. Buckner et al. (1981) also found that breeding for reduced grass tetany potential was possible in *Lolium-Festuca* hybrid derivatives, but they suggested that progress in those hybrids might be burdensome and that plants should be evaluated in widely differing environments. Reeder et al. (1986) found additionally that selection for high Mg concentration per se, as well as for low K/(Ca+Mg) ratio, would reduce the ratio because the two were negatively correlated. Increasing the ploidy level of tall fescue may increase its K/(Ca+Mg) ratio under some conditions (Kasperbauer et al., 1987).

Stable differences among commercial cultivars of perennial ryegrass for Mg concentration in the foliage indicated that it should be possible to select for a cultivar that has greater than normal Mg to alleviate grass tetany in grazing ruminants (Lane & Lamp, 1980). Currier et al. (1983) also found that heritability estimates of Mg and Ca concentration in orchardgrass were high and that breeding for improved Mg uptake in both flooded and nonflooded soil conditions should provide improved cultivars to alleviate grass tetany. Whereas Casler et al. (1987) found that smooth bromegrass herbage was severely deficient in seven minerals required by dairy cows when the grass was at advanced stages of maturation, and excessive in K/(Ca+Mg) at all stages, they suggested that these deficiencies could be more effectively corrected by fertilization of the soil or mineral supplementation of the forage during feeding than by breeding (even though some of the deficiencies could be corrected by breeding).

Crude Protein and Associated Traits

Hacker (1982) reviewed nine reports that generally showed negative correlations between crude protein (N) and water soluble carbohydrate concentration in temperate forage grasses, but there was sometimes sufficient genotypic independence to allow selection for both properties simultaneously. A varying relationship appeared between grass digestibility and N concentration, with the majority of correlations being positive. Lamb et al. (1984) showed that variability for digestibility and protein concentration in crested wheatgrasses [*Agropyron cristatum* (L.) Gaertn. spp. *pectinatum* (Bieb.) Tzvel. and *A. desertorum* (Fisch. ex Link) should allow breeding for both to improve feeding value. But, Vogel et al. (1986) reported that improving the protein concentration of intermediate wheatgrass by breeding might be difficult because of limited variability and large genotype × location effects for this trait, and Ehlke et al. (1986) found that CP concentration was rarely well correlated to IVDDM in divergent selections for IVDDM of smooth bromegrass.

Greub et al. (1986) found sufficient genotypic variation within a quackgrass population to allow breeding for N concentration as well as IVDDM, but total N was identified as the best selection criterion because of its better repeatability and lower genotype × year interaction.

In tall fescue, Soh et al. (1984) found that CP concentration was not correlated with IVDDM but at times was negatively associated with ADF and NDF, suggesting that simultaneous selection for higher CP and lower fiber fractions was possible. However, Kasperbauer et al. (1987) reported that increasing ploidy level in tall fescue may increase both CP and IVDDM concentrations.

Digestibility and Associated Chemical Components

Because intake of digestible energy is often the greatest factor that limits the performance of grass-fed ruminants, much attention has been given by plant breeders to selection for improved digestibility and reductions in the associated fiber fractions that may limit digestibility. When two grasses vary widely in digestibility due to maturation or environmental factors, digestibility and intake are often highly positively correlated. Concentration of ADF in temperate grasses is often most highly negatively correlated with digestibility, whereas NDF concentration is often most highly negatively correlated with intake (keeping in mind that concentrations of ADF and NDF are often positively correlated), as reviewed by Van Soest and Mertens (1977) and by Marten (1981a).

Two examples of increased animal performance response to increased digestibility per se via grass breeding are those documented by Burton et al. (1967) and Chapman et al. (1971) for 'Coastcross 1' bermudagrass (bred by Burton, 1972) and by Anderson et al. (1988) for 'Trailblazer' switchgrass (bred by Vogel et al., 1981, 1984). A 12% unit increase in digestibility (nylon bag estimate) of Coastcross over the bermudagrass cv. Coastal resulted in more than 30% increase in average daily gains by yearling steers and up to 50% increased steer gains per hectare. A 2% unit increase in IVDDM of total herbage of Trailblazer over the switchgrass cv. Pathfinder resulted in a 24% increase in both the average daily gains by grazing yearling cattle and animal liveweight gains per hectare (Table 6–2).

Hacker (1982) presented evidence for variation in digestibility and its heritability for a wide range of grasses in 21 studies where harvests within species had been taken on the same date and in nine studies where harvests had occurred at the same physiological age. Often, there was a range of at least 10 digestibility units, and about half of the variation was genetic. Both general and specific combining ability were important. When comparisons were at the same calendar date, late-flowering genotypes were more digestible; but, when comparisons were at the flowering stage, early flowering genotypes were usually more digestible.

Table 6–2. Digestibility of two switchgrass cultivars and performance of yearling cattle that grazed them during 3 yr in Nebraska. Adapted from Anderson et al. (1988).

Measurement	Cultivar	
	Trailblazer	Pathfinder
IVDDM (% dry wt.) of total available forage†	58.0	56.2
Average daily gain of cattle, kg/animal	0.73 (124%)	0.59 (100%)
Liveweight gain, kg/ha	351 (124%)	284 (100%)

† Means of weekly samples during three grazing seasons.

In the majority of cases reported before 1980, a negative correlation occurred between digestibility and yield of grasses (van Bogaert, 1977; Hacker, 1982). Van Bogaert (1977) repeatedly emphasized the negative correlations between quality trait improvement and dry matter yield in forage grasses, but Hacker (1982) suggested that the correlations between digestibility and yield were usually weak, and most authors found enough variation in both traits to allow digestibility to be improved without sacrificing yield.

In reports before 1980, strong negative correlations appeared between digestibility of grasses and lignin concentration, as well as between digestibility and either NDF or ADF concentration (Hacker, 1982). Additional evidence of these associations has appeared since 1980 (Table 6-3). Again, the

Table 6-3. Examples since 1980 of the potential for breeding forage grasses to improve their in vitro digestible dry matter (IVDDM) via either ruminal fermentation, cellulase digestion, or acid pepsin solubility.

Species	Source	Associated chemical component and/or other property¶	Association of component or property with digestibility
Smooth bromegrass	Casler, 1986†	ADL and cellulose	Negative
		Hemicellulose as % of CW	Negative
		CP, ash, and DM yield	Negligible
	Casler & Ehlke, 1986†	DM yield	Negligible
	Ehlke et al., 1986†	ADL	Negative
	Casler, 1987†	IVDDM of NDF, ADF, cellulose, and hemicellulose	Positive
	Casler et al., 1987†	Rate and lag time of CW-DDM	Positive
Tall fescue	Soh et al., 1984†	ADF, NDF	Negative
		Hemicellulose and CP	Negligible
Lolium-Festuca hybrids	Buckner et al., 1981§	WSC (parents only)	Positive
Orchardgrass	Saiga, 1983‡	--	--
Quackgrass	Greub et al., 1986†	NDF	Negative
		CP	Positive
Forage sorghum	Fritz et al., 1981†	PL	Negative
	Pedersen et al., 1982†	ADL, ADF, and NDF	Negative
	Tarumoto & Masaoka, 1983‡	--	--
	Cherney et al., 1986†	PL, NDF, and cellulose	Negative
Switchgrass	Vogel et al., 1981†	DM yield and CP	Negligible
	Vogel et al., 1984†	NDF	Negative
		PL, ADF, CP, and ash	Negligible
	Godshalk et al., 1986†	DM yield	Negligible
Intermediate wheatgrass	Vogel et al., 1986†	DM yield	Positive
		CP	Variable

† Ruminal fermentation.
‡ Cellulase digestion.
§ Acid pepsin solubility.
¶ PL = Permanganate lignin: ADL = Acid detergent lignin; NDF = Neutral detergent fiber; ADF = Acid detergent fiber; WSC = Water soluble carbohydrates; CP = Crude protein; IVDDM = In vitro digestible dry matter; CW = Cell wall; DM = Dry matter; DDM = Digestible dry matter.

above fiber fractions were frequently, but not always, negatively correlated with digestibility within several grass species. Also, DM yield was usually not strongly associated with digestibility in these grasses. Therefore, we may conclude in the case of most forage grasses that the fortune cookie message, "quantity is the enemy of quality," is not valid after all.[3]

Recent evidence supports Hacker's (1982) conclusion that genetic variation for digestibility exists for most forage grasses, and that plant breeders should usually direct their attention to selecting for improved digestibility of leaf tissue rather than total herbage.

Intake

Selection of forage grasses for animal intake potential is more complex, because various chemical and physical properties as well as animal factors can affect intake (Bath, 1987). Even though a positive relationship exists between digestibility and intake of forage grasses, and selecting for improved digestibility may also cause improved intake, cultivars may have different intake potential at the same digestibility (Hacker, 1982). Because the amount of NDF in a ruminant diet has a major influence on dry matter intake (Mertens, 1982), grass breeders should theoretically improve animal intake of genotypes by breeding for reduced NDF (Marum et al., 1979). As already reviewed (section on Morphology and Maturation), leaf percentage was the best of additional predictors (to add to digestibility per se) in multiple regressions to predict intake of diverse tropical grasses (Minson, 1980).

The use of the amount of electrical energy required to grind hay under standard conditions, as an assessment of fibrousness, was proposed by Chenost (1966) to estimate forage intake. His fibrousness index was highly correlated to organic matter digestibility ($r = -0.93$), to dry matter intake ($r = -0.90$), and to digestible organic matter intake (r = -0.94) of 25 hays (including 18 grasses) that ranged more than 28% units in digestibility and more than twofold in intake by sheep. Jones et al. (1973) found that the fibrousness index of Chenost was the best of five predictors of intake of 91 grasses (range of species and cultivars) grown in Wales ($r = -0.91$). Minson (1980) also reported that grinding energy (added to digestibility in a multiple regression) was highly correlated ($r = 0.85$) to voluntary intake of 30 feeds composed of leaf and stem fractions of five tropical grasses each cut at three growth stages. Grinding energy was somewhat superior to chemical composition for this purpose; the best chemical predictors of intake (to add to digestibility in multiple regressions) were lignin ($r = 0.79$), NDF ($r = 0.72$) and N ($r = 0.72$). Grinding energy has been used routinely in some European grass breeding programs to screen genotypes for increased quality (Dr. Christian Paul, West Germany, 1987, personal communication).

Wilson (1985) suggested that whereas a negative association exists between forage intake and cell wall concentration and plant cell wall resistance

---

[3] Message discovered in a fortune cookie at a Chinese restaurant by Dr. Christian Paul, grass breeder of Braunsweig, West Germany, during his 1987 visit to the USA to discuss breeding for forage quality.

to passage from the rumen, physical processes of mastication and rumination are probably more important than is microbial fermentation in comminution of the fiber. Therefore, measurements of stiffness, tensile strength, and fracture properties of fiber bundles, such as proposed by Vincent (1983), merit more attention. However, Hacker (1982) concluded that digestibility was probably the most readily measured trait that is closely associated to intake. He suggested that physical traits such as tensile strength and leaf flexibility may be suitable measures of intake potential for specific species, but are not likely to have general value.

Pest Resistance

Foliar diseases often cause reduced CP, soluble carbohydrates, digestibility, and palatability, as well as increased fiber components, in grasses (Hacker, 1982; Marten, 1985a). Insects can also have deleterious effects on CP, digestibility and other feeding value parameters of grasses, but sometimes insects cause increased leaf/stem ratios that enhance feeding value. Generally, resistance to fungal diseases is controlled by dominant genes (Hacker, 1982).

Perhaps the most significant fungal influence on forage feeding value discovered to date is that of the *Acremonium* spp. (fungal endophytes) that are responsible for tall fescue toxicoses and ryegrass staggers in grazing ruminants. The endophyte was associated with poor performance of animals that consumed tall fescue pastures, hay, and seed (Hoveland et al., 1984). Infection of grasses, including tall fescue, by endophytes was also associated with the accumulation of toxic alkaloids (Bush et al., 1982; Bacon et al., 1986). Seed treatments can control the tall fescue endophyte (Williams et al., 1984), but breeders are finding the release of fungal endophyte-free cultivars to be an easier solution to the problem.

More than 10 cultivars of tall fescue are being marketed in the USA that are fungal endophyte-free. Among those released are 'Johnstone', 'AU-Triumph', 'Missouri 96', 'Martin', 'Mosark', 'Safe', and 'Forager'. Whether these cultivars escaped from the presence of the *Acremonium* pathogen, or whether they have genetic resistance to the pathogen, has not been determined in most cases. Even though endophyte-free seed of older cultivars such as 'Kentucky 31' and 'Kenhy' (which have a history of fungal endophyte-infected stands) is now being marketed, genetic resistance is not the basis for this lack of infection.

Concern that use of fungal endophyte-free cultivars of tall fescue may lead to its devastation by insects has been supported by destruction of low-endophyte perennial ryegrass pastures by the Argentine stem weevil in New Zealand and by selective "greenbug" attack of low-endophyte tall fescue in laboratory tests at the Univ. of Kentucky (Crops and Soils, 1987). Kentucky researchers have suggested that plant selection for susceptibility to an endophyte that produces only desirable alkaloids, to provide insect resistance without toxicity to animals, may be a solution to this dichotomy.

Burner et al. (1983) found a high degree of genetic variation within a tall fescue population for water-soluble carbohydrate concentration, yield, and disease rating, which indicated potential for genetic improvement in all three traits. Pearl millet yield, DM, and IVDDM were all reduced as a result of susceptibility to rust (*Puccinia substriata* var. *indica*); infected genotypes had 51% less yield of DDM compared to resistant genotypes (Monson et al., 1986). They also reported that forage total sugars were considerably reduced, whereas CP concentration was slightly increased, by rust infection. Monson et al. (1986) also reviewed reports of negative effects on forage quality caused by rust infection of other species.

A leaf spot disease (*Cochliobolus sativus* Ito & Kurib.) of six strains of intermediate wheatgrass reduced mean forage digestibility by 5.5% units and increased NDF concentration by 4.7% units, whereas N level was not influenced by infection, in the glasshouse (Karn & Krupinsky, 1983). Differential responses to infection occurred for all quality traits among grass strains that showed potential for selection. In follow-up heritability studies, Berdahl and Krupinsky (1987) found that selection in intermediate wheatgrass for resistance to *C. sativus* infection would be effective only in field plots that had been established for at least a year. Karn and Krupinsky also reported that forage of field-grown plants of intermediate wheatgrass infected with stem smut (*Ustilago spegazzinii* Hirschh.) had reduced digestibility and increased concentrations of NDF, ADF, and lignin.

After three cycles of recurrent phenotypic selection in orchardgrass for resistance to purple leaf spot (caused by *Stagonospora arenaria* Sacc.), Obersheim et al. (1987) found significant differences among half-sib families for CP, IVDDM, ADF, and NDF concentrations, although they did not cause appreciable changes in these forage quality traits. They concluded that if higher forage quality is desired in combination with resistance to purple leaf spot in orchardgrass, it may be necessary to intermate the resistant population and populations that have superior quality. Little purple leaf spot occurred in the plots during their field nursery tests.

Recent evidence confirms Hacker's (1982) conclusion that improvement in grass quality traits via breeding for disease resistance should not cause undesirable genetically correlated responses. In cases where major disease infection is prevented in the field by breeding resistant cultivars, animal performance potential could be considerably enhanced.

Co-existence of Grasses with Legumes

The fact that legumes usually have more rapid dry matter and cell wall digestion, less cell wall (NDF) concentration, and greater intake potential than grasses will often lead pasture and hay-crop managers to attempt to include the maximum amount of legume possible in forage systems. The superior performance by cattle and sheep that have grazed legume-grass compared to N-fertilized, grass-based, pastures confirms the merit of legume-grass systems (Reed, 1981). For example, Burns and Standaert (1985) reported that in 31 experiments in the USA, steer or calf gains averaged from 0.14 to 0.15 kg/d more when the animals grazed legume-grass mixtures rather than all grass pastures.

The narrower range of adaptation of legumes, the threat of erosion, the often greater difficulty in legume stand establishment and maintenance, and the threat of ruminant bloat from many legumes all cause these same managers to select grasses or grass-legume mixtures for many less than ideal ecosystems or for practical management systems.

When the legume-grass mixture becomes the option of choice, and especially when a non-bloat-inducing legume (such as *Lotus corniculatus* L. or *Astragalus cicer* L.) is the pasture legume of choice, the manager may wish to have the grass present only to prevent erosion and/or the encroachment of low-quality weeds (especially perennial weeds). In these circumstances, a nondominant grass that will essentially maintain itself or spread only to fill yield-reducing vacancies left by dying legumes may be desired. The nutritional desirability of this option for a hay crop was well illustrated by the research of Reid and Templeton (1984) when they found that lambs appeared to regulate consumption of various combinations of temperate species grass and legume hay to constant neutral detergent fiber (cell wall) intake of 43 g/kg of body wt.$^{0.75}$ per day. Thus, the lambs consumed more digestible energy needed to maximize performance when the grass component of the mixtures was minimum.

Some grass breeders have recognized the desirability of evaluating their germplasm when it is mixed with legumes (especially alfalfa) to simulate practical production conditions. Reich and Casler (1985) reported that differences among nine polycross smooth bromegrass families were more frequently significant when the grass was grown in mixtures with alfalfa for NDF, ADF, and cellulose concentration, whereas differences were more frequently significant in pure grass stands for ADL. They found that with uniparental control of pollination, selection for improved forage quality in pure grass stands would not usually identify genotypes having improved quality in mixed grass-alfalfa stands. But, use of biparental control of pollination (to double selection gains) with selection at vegetative growth stages in pure stands would be more efficient than direct selection at an early heading target growth stage in grass-alfalfa mixtures for reducing NDF, ADF, and cellulose concentration in smooth bromegrass. They concluded that sampling of breeding populations at vegetative or early jointing stages would allow sufficient time for evaluation of forage quality (via NIRS) before anthesis. Then, biparental control of pollination could be used and selection for improved quality in pure stands might also improve grass quality in grass-alfalfa mixtures.

Jones (1987) defined the traits that were amenable to genetic selection that could improve compatibility of alfalfa or birdsfoot trefoil in binary mixtures with reed canarygrass. He found that height and early maturation of reed canarygrass were associated with its competitiveness with the legumes, and that tiller density (an important grass yield component) had a greater competitive effect than tiller weight. We have also observed in cattle-grazing experiments in progress that late-maturing timothy (*Phleum pratense* L.) cultivars are less competitive with birdsfoot trefoil than are early maturing cultivars. Jones (1987) concluded that selection of reed canarygrass for high dry matter yield, height, early maturation, and low rhizomatous spreading ability should improve its compatibility with legumes (in his study both alfalfa and birdsfoot trefoil dominated reed canarygrass following the establishment period). Because we have observed that reed canarygrass may dominate birdsfoot trefoil when they are established together on soils having a high N status, there would be merit in use of bidirectional (divergent) selection in reed canarygrass for the traits identified by Jones to provide diverse cultivars, some of which mix well with nondominant legumes and others that mix well with dominant legumes.

Other recent efforts have been directed toward development of late-maturing orchardgrass cultivars to enable their improved compatibility with legumes and greater feeding value of forage mixtures (Short & Carlson, 1987). These researchers found that less vigorous populations of orchardgrass permitted greater birdsfoot trefoil production in mixtures. They concluded that it is possible to increase the compatibility of orchardgrass and birdsfoot trefoil by selection among spaced plants of the grass for decreased spring canopy height, decreased tiller number, and delayed maturation.

Inherent Antiquality Traits and Palatability

Allelochemical Substances

Antiquality compounds that are synthesized by an organism that inhibits another organism may be classified as allelochemical substances. Crop Science Society of America Spec. Publ. 4 (Matches, 1973) included reviews of some of the allelochemicals in grasses that may adversely affect animal performance. Also, Hacker (1982) reviewed evidence of genetic variation within or among grass species for cyanogenic glycosides, nitrates, alkaloids, tannins, and oxalates. Among other recent reports and reviews of genetic variability in forage grasses for antiquality compounds are those of Haskins et al. (1987) for hydrocyanic acid potential in sorghums; Oram et al. (1985) for alkaloids in *Phalaris aquatica* L.; Marten (1981b) for alkaloids in several species and cyanogenic glycosides in sorghums; Bacon et al. (1986) for alkaloids in several grass species; Sleper (1985) for alkaloids in tall fescue; Marum et al. (1979a) for alkaloids in reed canarygrass; and Hovin and Marten (1983) for alkaloids in reed canarygrass.

According to Yates (S.G. Yates, 1987, personal communication), three crucial preliminary steps must be accomplished before plant breeders can confidently proceed in heritability and selection programs to reduce antiquality constituents in forages: (i) isolation, (ii) characterization, and (iii) analysis-quantification of the allelochemical(s).

Palatability Relative to Chemical and Physical Traits

An example of progress in breeding to reduce an antiquality factor in a grass that has had commercial significance is the reduction of indole alkaloids in reed canarygrass. The research program that identified the role of indole alkaloids as palatability inhibitors and animal performance depressants in reed canarygrass and the results of breeding to modify their type and reduce their quantity is outlined in a recent book chapter (Marten, 1985b). The primary findings in this series of experiments were as follows:

1. The lack of palatability of 'Rise' reed canarygrass was associated with lower average daily gains by lambs that grazed it without choice compared to more palatable smooth bromegrass or orchardgrass; lambs exhibited diarrhea only when they grazed reed canarygrass.

2. Palatability ratings of more than 400 genotypes of reed canarygrass from the world collection were negatively correlated with their total indole alkaloid concentration; the threshold level of alkaloids beyond which sheep consistently rejected specific genotypes was about 0.6% of the dry matter, but sometimes rejection started at 0.4% of the dry matter.

3. A positive association occurred between absolute intake (when a choice was not offered) by sheep and palatability of reed canarygrass genotypes. Indole alkaloid concentration of eight genotypes was negatively correlated with intake by sheep; however, intake by steers was not correlated with alkaloid concentration.

4. Total indole alkaloid concentration of eight diverse genotypes of reed canarygrass was highly negatively associated with average daily gains by both lambs and steers confined to pastures of individual genotypes (Table 6–4). Both classes of animals had more diarrhea when they grazed high-alkaloid plants and when they grazed tryptamine-carboline-containing compared to gramine-containing plants.

5. Lambs that grazed the low-alkaloid (gramine only) MN-76 germplasm selection of reed canarygrass gained from 51 to 95% more per day than those that grazed the variety Rise (mixed alkaloid types). MN-76 contained 0.09 to 0.12% alkaloid, whereas Rise contained 0.28 to 0.32% alkaloid. The correlation (r) between total indole alkaloid concentration of these and other entries in the experiment and lamb daily gain ranged from -0.66 to -0.97 (Table 6–5). We concluded that the threshold level of alkaloids in reed canarygrass beyond which lambs will show reduced gain is about 0.2% of the dry matter (one-third that required for sheep to consistently reject genotypes in palatability trials).

Table 6–4. Average daily gains by lambs or young steers grazing reed canarygrass that had high vs. low concentrations of indole alkaloids.

	Lamb gains		Steer gains
Alkaloid level	Yr 1	Yr 2	Yr 3
	g/animal per d		
High (0.31% dry wt.)	−1b*	71b	140b
Low (0.10% dry wt.)	82a	118a	330a

* Different letters following pairs of figures within years indicate statistical difference at $P \leq 0.05$. Adapted from Marten et al. (1976).

Table 6–5. Average daily gains and incidence of diarrhea by lambs grazing MN-76 vs. Rise reed canarygrass.

Selection	Lamb gains		Diarrhea incidence	
	Yr 1	Yr 2	Yr 1	Yr 2
	g/d		%	
Rise (0.30% dry wt. as alkaloid)	64b*	71b	26a	13a
MN-76 (0.10% dry wt. as alkaloid)	125a	107a	1b	2b

* Different letters following pairs of figures within years indicate statistical difference at $P \leq 0.05$. Adapted from Marten et al. (1981).

6. Lambs gained as much or more when they grazed low alkaloid (MN-76) reed canarygrass compared to smooth bromegrass. This documented that the breeding program to reduce total indole alkaloid concentration and to remove tryptamine-carboline alkaloids completely from reed canarygrass led to a biologically significant improvement in the species (Hovin & Marten, 1983). Since then, industry breeders in Iowa have released the cv. Palaton and Venture that share the low-alkaloid characteristics of MN-76, and a Canadian cv. (Rival) that has low alkaloids has been released.

The results of the above series of experiments appeared to endorse Voigt's (1975) conclusion that breeding for palatability is worthwhile if antiquality constituents are associated with the palatability problem.

An example of a small improvement in animal performance as the result of breeding for palatability improvement even when no known allelochemical was involved is that of Voigt et al. (1970). They bred the more palatable selection, '944' ('Morpa'), of weeping lovegrass [*Eragrostis curvula* (Schrad.) Nees.] that improved daily gains ($P \leq 0.05$) in grazing steers during both winter and summer (3-yr means of 0.49, 0.44, and 0.44 kg/animal per day for the palatable 944, unpalatable common, and unpalatable '673', respectively). The greater palatability of the improved cultivar was also associated with reduced lignin, increased cellulose, and reduced lignin/cellulose ratio. The palatable selection was also taller, later heading, and had slightly wider leaves than the unpalatable entries.

Sleper (1985) suggested that an important factor in breeding tall fescue for greater animal performance potential would be to improve its palatability. Tall fescue is among the least palatable grasses adapted to northern USA. Voigt (1975), Hacker (1982), Sleper (1985) and Nguyen et al. (1982) reviewed research in Europe, South Africa, New Zealand, the USA, and Japan in which either tall fescue, orchardgrass, ryegrass, weeping lovegrass, or other grass species genotypes were selected for either leaf flexibility, leaf tensile strength, reduced alkaloids, soluble carbohydrates, or palatability per se in attempts to improve their intake. Whereas some progress was reported, apparently no substantial evidence of improvement of animal performance as a result of palatability improvement of these species has appeared (with the exception of the report of Voigt et al., cited above). When we tested an orchardgrass cultivar bred in the Netherlands for reduced presence of "silicious dentations'' (silica barbs) on the leaf blade margins, grazing sheep could not distinguish this cultivar from other cultivars that had silica barbs (G.C. Marten, 1979, unpublished data).

Palatability variation in *Lolium* × *Festuca* hybrids has been investigated by plant breeders who have attempted to incorporate the greater palatability of ryegrass into unpalatable tall fescue germplasm. Berg et al. (1979) found that synthetics from palatable parents had greater IVDDM than those from unpalatable parents. Buckner et al. (1979) reported that when succulence (water concentration) was used as a hybrid derivative selection index during periods of moisture stress, a positive association existed among water concentration (reduced leaf roll), palatability, and digestibility. Water concentration was negatively correlated with ADF. Water concentration under these conditions was sufficiently heritable to allow its enhancement via breeding.

Hanna et al. (1974) reported that a single gene mutant of pearl millet, which suppresses the formation of shoot trichomes and which has a more waxy surface than its isogenic hairy counterpart, was more palatable to grazing ruminants but less digestible in vitro. Thus, breeding for nonhairy cultivars of pearl millet to improve palatability is likely to have a negative effect on digestibility. Burton (1974) reported that Coastal bermudagrass was more palatable to dairy cows than was a common cultivar, and that average daily gains of steers were 16% better (5-yr mean) when they grazed Coastal compared to common bermudagrass. He concluded that either the steers consumed more of the Coastal or that it may have been more nutritious. However, Burton also concluded from other experiments that differences in palatability of cultivars within species in cafeteria trials may have only limited value in predicting intake or animal performance when no choice is offered.

LABORATORY METHODS FOR GENOTYPE SELECTION TO MAXIMIZE GRASS FEEDING VALUE

Numerous laboratory approaches have been suggested, but relatively few have been routinely used by plant breeders to screen germplasm for improved digestible energy and intake potential. Several comprehensive reviews (outlined below) have been published on this topic, and the reader is referred to these reviews for details. Because of a lack of accurate, fast, and inexpensive methods to screen for improved feeding value of grasses, most breeders did not initiate such programs on an extensive scale until after 1960 (van Bogaert, 1977).

Analytical Approaches

Among laboratory methods for predicting forage quality that have been developed during the past 35 yr are those classified as definitive chemical, solubility indices (including cellulase enzyme), artificial mastication, histological, fibrousness index (grinding energy), nylon or dacron bag, in vivo artificial rumen, and in vitro rumen fermentation by Barnes (1973). His review was updated to include new developments in 1979 (Barnes & Marten, 1979). A comprehensive assessment of the utility of in vitro rumen fermentation and fungal enzyme (cellulase) systems, and a suggested standardized in vitro procedure developed by the NC-64 Regional Committee, were published by Marten and Barnes (1980).

Marten (1981a) and Minson (1981a) presented U.S. and Australian views of the merit of various laboratory techniques for evaluating forage quality. They agreed that when forages having wide ranges in digestibility and intake were included in prediction equations, high correlations could be easily found between these quality traits and numerous chemical and/or physical measurements. However, some of these measurements were of limited value to plant breeders who must select within narrow ranges of maturation or within plant parts. Marten and Minson also agreed that digestibility measured by a reliable in vitro rumen fermentation or cellulase approach was a superior selection criterion to recommend to breeders who wish to improve animal performance potential of grasses. Marten further supported use of cell wall (NDF) concentration and Minson supported use of leafiness as valuable methods of screening for improved forage quality.

Recent additional evidence of the value of potent fungal cellulase solutions for screening grass germplasm for improved digestibility is that of Sleper and Roughan (1984), Bughrara and Sleper (1986), Gabrielsen (1986), and Marten et al. (1988). Marten et al. (1988) found that wavelength selection during NIRS equation development appeared to be based on chemical entities that were often commonly digested by rumen microorganisms and cultured cellulase; they also reported that IVDDM and fungal cellulase methods usually ranked the digestibility of grass and other forages similarly, and that NIRS was successfully used to estimate digestibility by either in vitro method.

Casler (1986) used five independent chemical variables in a path analysis model that detailed hypothesized causal effects and correlated causes among forage constituents, IVDDM, forage yield, and residual effects in smooth bromegrass. He concluded that a cell wall ideotype for improved digestibility of this species should have reduced cellulose and lignin and either reduced or unchanged hemicellulose concentration. All three of these cell wall fractions could be estimated by NIRS for efficient breeding progress.

Near Infrared Reflectance Spectroscopy

The potential, history of development, and initial successful use of NIRS for evaluating the feeding value of grasses and other forages are reviewed in USDA Agric. Handb. 643 (Marten et al., 1985). Shenk (1977) concluded that NIRS may remove the barriers to forage quality evaluation by plant breeders that have existed in the past. Marten (1985c) stated that because the typical plant breeder seeks relative rather than absolute forage quality values, bias concerns are kept to a minimum. The breeder is most concerned about the precise measurement of thousands of plants. Usually, only small amounts of forage are required by NIRS, and this requirement facilitates quality testing of single plants or even portions of single plants. The breeder will also be interested in knowing which plants are outliers to the sample population. Available NIRS computer programs will identify outliers. The breeder can then determine by conventional assay whether the outlier sample has been inaccurately assessed by NIRS (indicating the need for its addition to the calibration set) and/or whether it has a truly unique forage quality. Because NIRS can easily provide the fast, precise, and outlier information sought by the plant breeder, plant breeding is the uppermost area of research for which it can be recommended.

Marum et al. (1979b) reported the first use of NIRS to measure forage quality (six constituents) in a genetics and breeding program. They determined heritability of cell wall constituents and associated quality components in reed canarygrass. Since then, numerous grass breeding programs based on use of NIRS have been initiated worldwide. Among recent examples of successful use of NIRS in grass breeding and genetic studies are the smooth bromegrass reports of Reich and Casler (1985) and Ehlke et al. (1986), and the bermudagrass report of Burton et al. (1987). Also, de Ruiter and Burns (1987d) found NIRS gave good analysis of the concentration of carbohydrates in partially fermented cell walls, as well as the extent of dry matter loss during fermentation of flaccidgrass.

Casler (1985) concluded that because of the ease and speed of NIRS compared to conventional laboratory assays, one can generate larger populations and greater selection intensities in grass quality breeding programs, improve pollination control by selection before anthesis in the year of evaluation, shorten cycle times for recurrent selection, and identify appropriate selection criteria via NIRS.

NEW HORIZONS IN BREEDING FORAGE GRASSES TO MAXIMIZE ANIMAL PERFORMANCE

A strong case can be made for the wisdom of continuing the efforts to select grass genotypes to improve their digestibility and intake and to remove or reduce antiquality constituents using the best of the screening methods identified above. Scientists at the CSIRO-DSIR Workshop on Plant Breeding and Feeding Value of Pastures and Forages (Ulyatt & Black, 1984) agreed that in cases where satisfactory progress has been made on species persistence and disease problems, there are special opportunities to select for feeding value. Selection for feeding value could be based on agronomic characteristics such as leafiness and extension of the growth period of vegetative tillers. There is potential, however, for selecting for specific feeding value characters, including voluntary intake (through increased digestibility of fiber or rate of fiber outflow), decreased protein degradability in the rumen, efficiency of nutrient utilization (through increased soluble carbohydrate or lipid and reduced lignocellulose concentrations), reduced toxic factors, and increased ingestibility (through increased palatability or improved sward profile).

Plant breeders have historically awaited the development of new evidence that a forage quality parameter was biologically significant to animal performance or that a laboratory method was highly correlated with digestibility or an antiquality trait before selection of improved germplasm was undertaken. Then, in many instances, selection was "unidirectional" toward development of an improved cultivar that could be tested in grazing or confined feeding experiments to determine whether animal performance improvement would result. This logical and apparently prudent approach has led to some decided successes, some of which are described earlier. Continued endeavors of this sort are certainly not to be discouraged, but I suggest that we must be bolder and more imaginative if plant breeders are to efficiently exploit new knowledge concerning forage quality as it is being developed.

What we need in the future to make more rapid progress in breeding forage grasses to maximize animal performance is a "*bidirectional*," "*divergent*," or "*differential*" *selection approach* based on *theoretical* as well as on proven avenues to forage quality improvement. If the plant breeder is willing to exert the additional effort to select for *both low and high quality* for whatever trait(s) are being pursued, the task of the forage utilization scientists who determine the biological significance of the selections will be made easier and the mechanisms of the animal reaction will be more easily determined.

Recent reports reveal some of the theoretical approaches that may lead to major advances in the performance of animals that consume forages. Among these are the reports given at the Tri-divisional (Crop Science Society of America) symposium entitled, "Histological and Physical Factors Influencing Forage Digestibility and Intake" at this 79th Annual Meeting of the American Society of Agronomy. Refer to *Agron. J.* 81:17–46 for the four reports presented by D.E. Akin (p. 17–25); D.S. Fisher et al. (p. 25–33); H.G. Jung (p. 33–38); and R.D. Hatfield (p. 39–46). The presiding officer (D.R. Buxton, 1987, personal communication) has summarized the ideas emanating from this symposium relative to selection criteria that might be worthwhile for forage grass breeders as follows:

1. Select for altered composition of structural carbohydrates including higher levels of pectic polysaccharides and lower levels of cellulose and hemicellulose.
2. Select for altered composition of noncore and core lignin so that the digestion inhibition by lignin is reduced.
3. Select for a higher proportion of cell and tissue types that are digestible within plant parts.
4. Select for a cuticle that imposes less resistance to microbial penetration.
5. Select for a fast rate of cell wall digestion.

Akin (1986) identified factors "indigenous to the plants per se", as well as high growth temperature, that contribute to the reduced nutritive value of warm-season grasses. The three categories of factors that are potentially amenable to genetic selection for improvement were: (i) structural carbohydrates, (ii) phenolic compounds, and (iii) plant anatomy. All three of these are represented in Buxton's above list for both warm- and cool-season forages.

Structural Carbohydrates

Within the structural carbohydrate category, Akin (1986) discounted cellulose crystallinity as a limitor of cellulose digestion; instead, he reviewed research that suggested that the organization of the cell wall (rather than composition), lignin interaction with hemicellulose, and deposits of cutin, silica, and other minerals that protect structural carbohydrates from microbial attack are worthy of consideration. Perhaps investigations discussed by Hatfield (1987) and of rates or extent of degradation of individual polysaccharides by rumen microorganisms will give some clues that are useful to the grass breeder.

Recent comparisons of the digestible and indigestible cell wall carbohydrates of flaccidgrass (de Ruiter & Burns, 1987a, b, c) indicated that specific cell wall sugars (especially xylose and glucose) may be strongly negatively associated with DM digestibility. Also, Buxton et al. (1987) reported that glucose comprised more than 60% of the total neutral sugars in the cell wall polysaccharides of smooth bromegrass and orchardgrass stems, whereas xylose comprised 30%. However, in spite of large differences in proportions of neutral sugars among grass and legume species, the only sugar associated with in vitro digestibility (independent of lignin) was mannose in grass stems.

Phenolic Compounds

According to reviews cited by Akin (1986), the phenylpropanoid polymer lignin is the most widely recognized limitation to the extent of digestibility of cell walls and forage DM. He suggested that if p-coumaryl moieties in esters or within condensed lignin are the phenolic groups best able to bind cell wall components, this may explain why tissues in the tropical bermudagrass are less attacked by delignifying agents and are less degraded than tissues in the temperate tall fescue. Variation in lignin types between grass species are also shown by histological staining.

Jung and Fahey (1983) suggested that low molecular weight phenolic compounds may be responsible for poor digestion of some forages. Jung (1987) pointed out that even though polymeric core lignin has received the major emphasis relative to impediment of forage cell wall digestibility during the last 50 yr, he and his colleagues have been focusing on noncore lignin. The primary noncore lignin components are p-coumaric and ferulic acids that are esterified to core lignin and hemicellulose. Noncore lignins occur in greater concentrations in grasses than in legumes (the reverse of the situation for core lignin). Jung (1987) reported that both genetic and environmental factors influence the concentration and composition of lignins in grasses and other forages. Whereas fiber fermentation per se may be more inhibited by core lignin, noncore lignin depresses fermentative activity of rumen microorganisms.

Research reviewed by Akin (1986) revealed that for both warm- and cool-season grasses, p-coumaric acid has been more negatively correlated to digestibility than has ferulic acid. Also, warm-season grasses have a greater concentration of p-coumaric acid, as well as a higher ratio of p-coumaric/ferulic acid, than do cool-season grasses. Increased levels of both acids in tall fescue at warm temperatures indicated that these compounds may contribute to lower digestibility of leaf blades when plants are grown at warm temperatures (Akin et al., 1987).

Cherney et al. (1986) reported a higher concentration of cell wall ferulic acid and lower p-coumaric acid in brown-midrib sorghum than in less digestible normal sorghum; this indicated potential differences in linkages between core lignin and structural carbohydrates in the two sorghum types. Akin et al. (1986b) also found that reduced concentrations of phenolic acids (especially p-coumaric acid) and reduced etherated syringyl moieties occurred in association with reduced lignin of the more digestible brown midrib-12 sorghum.

If research reveals which lignin components are most limiting to grass fiber digestion, plant breeders may be able to reduce the influence of these components via selection. Breeding to modify the biosynthesis of phenolic compounds in grass cell walls could produce plants that have improved rates of fiber digestion. Indeed, Vogel and Gabrielsen in Nebraska (1987, personal communication) have initiated a breeding program to modify phenolic ratios in switchgrass. They found that in selecting for improved IVDDM they also inadvertently changed the ratio of p-coumaric/ferulic acids in this species, without changing the concentrations of NDF, ADF, or lignin (Gabrielsen et al., 1987). They increased the concentration of ferulic acid, decreased the concentration of p-coumaric acid, and decreased the p-coumaric/ferulic acid ratio in the cv. Trailblazer by increasing its IVDDM, which was seven percentage units greater than that of a low IVDDM strain. The greater yearling cattle gains reported by Anderson et al. (1988), and discussed under the section on "digestibility and associated chemical components," for Trailblazer switchgrass compared to Pathfinder were also associated with reduced p-coumaric acid concentration and a lower p-coumaric/ferulic acid ratio for Trailblazer.

Plant Anatomy

Much research has revealed basic anatomical differences that may influence digestibility between warm- and cool-season grasses (Akin, 1986). Most warm-season (C_4) types have a Kranz leaf anatomy that results in closely spaced vascular bundles and a distinct, thick-walled parenchyma bundle sheath surrounding the bundles (Fig. 6–2). Cool-season grasses are C_3 types that have widely spaced bundles and less pronounced parenchyma bundle sheaths. The loose arrangement of the bundles in the C_3 grasses allows easier penetration of rumen microorganisms. Also, the bundle sheaths in the C_4 grasses are slowly or only partially degraded (to lead to a rigid residue) compared to rapid and extensive degradation of bundle sheaths in C_3 grasses. In some C_4 species, phenolic compounds appear to limit bundle sheath degradation and a suberin structure may limit cell wall digestion.

Akin et al. (1986a) reported that reduced lignin concentration in brown midrib mutants of C_4 sorghum (associated with improved digestibility) was not associated with leaf blade anatomy or histological reactions for lignin and did not change microbial degradation of the most resistant tissues. The mutants did have improved microbial degradation of those tissues that are usually only poorly degraded. They also found (Akin et al., 1986b) that degradation of delignified midribs in brown midrib mutants by rumen microbes indicated that these tissues were, indeed, more degradable than those of the midribs of normal sorghum.

Wilson et al. (1983) reported that in addition to the gross differences in tissue proportions between Kranz and non-Kranz anatomy shown by less digestible C_4 and more digestible C_3 *Panicum* spp., leaf bulk density (dry wt./leaf volume) was negatively correlated with digestibility. Akin et al. (1983) concluded that improved *Panicum* digestibility could result from breeding to increase the degradation of the parenchyma bundle sheaths of the C_4 species, without changing the basic leaf anatomical structure.

Hacker (1982) reviewed reports of associations as well as lack of associations between digestibility and anatomical characteristics within perennial ryegrass, Kentucky bluegrass (*Poa pratensis* L.), smooth bromegrass, limpograss [*Hemarthria altissima* (Poiret) Stapf & C.E. Hubb.], and bermudagrass. Ehlke and Casler (1985) also reported that highly digestible smooth bromegrass clones had greater percentages of readily digestible stem chlorenchyma and leaf mesophyll, as well as lesser amounts of lignified leaf vascular tissue and thicker stems, than did less well-digested clones. Reduced rates of digestibility by reed canarygrass compared to a perennial ryegrass hybrid were related to differences in leaf anatomy that allowed the ryegrass leaves to be more readily penetrated by rumen microbes (Grabber & Allinson, 1985).

Akin (1986) further reported that the epidermal cuticle is usually not degraded by rumen microorganisms in either warm- or cool-season grasses. He also reviewed other anatomical features of grass leaves and stems that could influence their digestibility.

Because of the lack of consistency in reports of relationships between anatomical traits and digestibility *within* either cool- or warm-season grass species, a generalized screening for anatomical traits by breeders cannot be recommended. Methods that may be useful in the quest for germplasm having specific anatomical properties include the "section to slide techniques" (Akin, 1982) and the "image analysis" procedure of Twidwell et al. (1987). As specific associations are revealed within particular species, attempts should be made to quantify anatomical properties in tissues that can be dried and ground to provide calibration sets, which may then allow use of NIRS to screen large populations for improved digestibility based on improved anatomy.

Physical Traits and Intake

Akin (1986) cited research that illustrated how the rigid residue after microbial degradation in stems could be a physical factor causing fiber retention in the rumen and resulting in reduced intake of stems compared to that of leaves. Additionally, he reviewed experiments that revealed variation in the physical breakdown by mastication of C_4 and C_3 grass leaves that could relate to intake as well as digestibility. Thus, selection for a higher proportion of cell and tissue types within leaves or stems of grasses for ease of mastication might improve both digestibility and intake.

The speed of particle flow through the ruminant digestive tract may often determine intake potential of a grass. If plant breeders can change anatomical structures in grasses relative to physical breakdown and flow of undigested material, rates of fiber passage may increase to enhance intake (Welch, 1986). Of special concern are particle size and particle specific gravity (small particles of high specific gravity pass quickly from the rumen). Akin (1987, personal communication) suggested that stem anatomy may be more manipulable than that of leaves relative to particle passage phenomena.

Wilson (1985) agreed that the concept of large and small particle pools and the processes that control particle escape from the rumen are germane to consideration of plant tissue anatomical structures and their susceptibility to fragmentation. He suggested that histological studies were needed relative to the effect of size, shape, and density of particles arising from tissues. Plant breeders are advised to carefully follow research in this area for new clues as they arise regarding possibilities for improving intake of grasses.

Protein Fractions

Recent discussions concerning the need to focus on protein quality as well as "crude protein" in the ruminant diet, especially relative to rumen escape or metabolizable protein, may lead to new research discoveries concerning protein properties of the nutritionally ideal grass. Excessive protein degradability in the rumen, accompanied by N loss as ammonia, apparently occurs during use of many forage grasses. New Zealand and Australian researchers have suggested that future investigations should include methods of decreasing the degradability of protein in grasses by incorporation of condensed tannin or by incorporation of insoluble proteins into herbage via genetic engineering (Ulyatt & Black, 1984).

Other Nonconventional Measurements to Consider or Reconsider

Perhaps we need to reconsider the merits of some of the theoretical physical methods of quality measurement that seem to have come and gone. Some of these (artificial mastication, fibrousness measured by grinding energy requirements, density indices, stiffness, tensile strength of leaves, etc.) seem to relate to the mechanical stability of plant tissues that could affect intake potential. For example, Paul et al. (1985) recommended a modified grinding energy test for routine screening of German grasses. Whereas these types of measurements are not easily standardized and are likely species specific as to their applicability to feeding value assessment, they may provide new avenues to quality improvement.

If a specific physical trait can be shown to be heritable in a grass species, and if plants with a reasonable range of genetic variation for the trait can be assembled, one should likely be able to develop a calibration set to enable routine screening for divergent plants via NIRS. This is true because most such physical traits will likely be conditioned by some (perhaps unknown) chemical bonding that can be detected by differential absorption of NIR radiation.

Then, if plant breeders are willing to expend the effort to develop divergent strains of a grass and solicit interdisciplinary efforts to enable carefully controlled grazing studies or confined feeding experiments with large numbers of ruminant animals, we may make great progress in forage quality improvement by "coming in the back door". We have perhaps spent too much time waiting for ruminant nutritionists to "invite us through the front door" with carefully researched documentation of the nutritional meaning of a trait or concept before being willing to initiate grass breeding programs.

REFERENCES

Akin, D.E. 1982. Section to slide technique for study of forage anatomy and digestion. Crop Sci. 22:444–446.

Akin, D.E. 1983. Structural characteristics limiting digestion of forage fiber. p. 511–514. *In* J.A. Smith and V.W. Hays (ed.) Proc. 14th Int. Grassl. Congr. Westview Press, Boulder, CO.

Akin, D.E. 1986. Interaction of ruminal bacteria and fungi with southern forages. J. Anim. Sci. 63:962–977.

Akin, D.E., R.H. Brown, and L.L. Rigsby. 1984. Digestion of stem tissues in *Panicum* species. Crop Sci. 24:769–773.

Akin, D.E., and D. Burdick. 1975. Percentage of tissue types in tropical and temperate grass leaf blades and degradation of tissue by rumen microorganisms. Crop Sci. 15:661–668.

Akin, D.E., S.L. Fales, L.L. Rigsby, and M.E. Snook. 1987. Temperature effects on leaf anatomy, phenolic acids, and tissue digestibility in tall fescue. Agron. J. 79:271–275.

Akin, D.E., W.W. Hanna, and L.L. Rigsby. 1986a. Normal-12 and brown midrib-12 sorghum. I. Variations in tissue digestibility. Agron. J. 78:827–832.

Akin, D.E., W.W. Hanna, M.E. Snook, D.S. Himmelsbach, F.E. Barton, II, and W.R. Windham. 1986b. Normal-12 and brown midrib-12 sorghum. II. Chemical variations and digestibility. Agron. J. 78:832–837.

Akin, D.E., E.L. Robinson, F.E. Barton, II, and D.S. Himmelsbach. 1977. Changes with maturity in anatomy, histochemistry, chemistry and tissue digestibility of bermudagrass plant parts. J. Agric. Food Chem. 25:179–186.

Akin, D.E., J.R. Wilson, and W.R. Windham. 1983. Site and rate of tissue digestion of C_3, C_4, and C_3/C_4 intermediate *Panicum* species. Crop Sci. 23:147–155.

Anderson, B., J.K. Ward, K.P. Vogel, M.G. Ward, H.J. Gorz, and F.A. Haskins. 1988. Forage quality and performance of yearlings grazing switchgrass strains selected for differing digestibility. J. Anim. Sci. 66:2239–2244.

Anonymous. 1987. Low endophyte fescue runs into a snag. Crop Soils 39(6):26.

Asay, K.H. 1976. Breeding to increase quality in temperate grasses. p. 104. *In* Agronomy abstract. ASA, Madison, WI.

Bacon, C.W., P.C. Lyons, J.K. Porter, and J.D. Robbins. 1986. Ergot toxicity from endophyte-infected grasses: A review. Agron. J. 78:106–116.

Barnes, R.F. 1973. Laboratory methods of evaluating feeding value of herbage. p. 179–214. *In* G.W. Butler and R.W. Bailey (ed.) Chemistry and biochemistry of herbage. Vol. 3. Academic Press, New York.

Barnes, R.F., and G.C. Marten. 1979. Recent developments in predicting forage quality. J. Anim. Sci. 48:1554–1561.

Barton, F.E., II, H.E. Amos, D. Burdick, and R.L. Wilson. 1976. Relationship of chemical analysis to *in vitro* digestibility for selected tropical and temperate grasses. J. Anim. Sci. 43:504–512.

Bath, D.L. 1987. Dietary influences on feed intake in ruminants. p. 9–12. *In* Proc. 8th Western Nutr. Conf., Edmonton, Alberta. 24–25 Sept. Univ. of Alberta, Edmonton, Canada.

Berdahl, J.D., and J.M. Krupinsky. 1987. Heritability of resistance to leaf spot diseases in intermediate wheatgrass. Crop Sci. 27:5–8.

Berg, C.C., R.R. Hill, Jr., R.C. Buckner, and R.F Barnes. 1979. Forage production and quality of synthetics derived from *Lolium* × *Festuca* hybrids. Crop Sci. 19:89–93.

Berg, C.C., G.M. Wood, A.W. Hovin, and C.M. Rincker. 1981. Floral induction and forage yield and quality of sparse-flowering orchardgrass synthetics. Crop Sci. 21:677–680.

Buckner, R.C., P.B. Burrus, II, P.L. Cornelius, L.P. Bush, and J.E. Leggett. 1981. Genetic variability and heritability of certain forage quality and mineral constituents in *Lolium-Festuca* hybrid derivatives. Crop Sci. 21:419–423.

Buckner, R.C., L.P. Bush, and P.B. Burrus, II. 1979. Succulence as a selection criterion for improved forage quality in *Lolium-Festuca* hybrids. Crop Sci. 19:93–96.

Bughrara, S.S., and D.A. Sleper. 1986. Digestion of several temperate forage species by a prepared cellulase solution. Agron. J. 78:94–98.

Burner, D.M., J.A. Balasko, and W.V. Thayne. 1983. Genetic and environmental variance of water soluble carbohydrate concentration, yield, and disease in tall fescue. Crop Sci. 23:760–763.

Burns, J.C., and J.E. Standaert. 1985. Productivity and economics of legume-based vs. nitrogen fertilized grass-based pastures in the United States. p. 56–71. *In* R. F Barnes et al. (ed.) Proc. of a Trilateral Workshop, Forage legumes for energy-efficient animal production (USA-Australia-New Zealand). USDA-ARS, CSIRO, and DSIR. Natl. Tech. Info. Serv., Springfield, VA.

Burton, G.W. 1972. Registration of Coastcross-1 bermudagrass. Crop Sci. 12:125.

Burton, G.W. 1974. Improving forage quality by breeding. p. 705–714. *In* Proc. 12th Int. Grassl. Congr., Moscow, USSR. Vol. 3. Four Continent Book Corp., New York.

Burton, G.W. 1976. Breeding to improve quality in tropical grasses. p. 105. *In* Agronomy abstract. ASA, Madison, WI.

Burton, G.W. 1986. Developing better forages for the South. J. Anim. Sci. 63:955–961.

Burton, G.W., F.E. Barton, and W.G. Monson. 1987. NIRS vs. IVDMD methods to evaluate bermudagrass digestibility. p. 00. *In* Abstract, Southern Section of ASA Meet., Nashville, TN. ASA, Madison, WI.

Burton, G.W., R.H. Hart, and R.S. Lowrey. 1967. Improving forage quality in bermudagrass by breeding. Crop Sci. 7:329–332.

Burton, G.W., W.G. Monson, J.C. Johnson, R.S. Lowrey, H.D. Chapman, and W.H. Marchant. 1969. Effect of the d_2 dwarf gene on the forage yield and quality of pearl millet. Agron. J. 61:607–612.

Bush, L.P., P.L. Cornelius, R.C. Buckner, D.R. Varney, R.A. Chapman, P.B. Burrus, II, C.W. Kennedy, T.A. Jones, and M.J. Saunders. 1982. Association of N-acetyl loline and N-formyl loline with *Epichloe typhina* in tall fescue. Crop Sci. 22:941–943.

Buxton, D.R., J.R. Russell, and W.F. Wedin. 1987. Structural neutral sugars in legume and grass stems in relation to digestibility. Crop Sci. 27:1279–1285.

Carlson, I.T., D.K. Christensen, and R.B. Pearce. 1983. Selection for specific leaf weight in reed canarygrass and its effects on the plant. p. 207–209. *In* J.A. Smith and V.W. Hays (ed.) Proc. 14th Int. Grassl. Congr., Lexington, KY. 15–24 June 1981, Westview Press, Boulder, CO.

Casler, M.D. 1985. Breeding for improved forage nutritive value using NIRS. p. 125. *In* Agronomy abstract. ASA, Madison, WI.

Casler, M.D. 1986. Causal effects among forage yield and quality measures of smooth bromegrass. Can. J. Plant Sci. 66:591–600.

Casler, M.D. 1987. In vitro digestibility of dry matter and cell wall constituents in smooth bromegrass forage. Crop Sci. 27:931–934.

Casler, M.D., M. Collins, and J.M. Reich. 1987. Location, year, maturity, and alfalfa competition effects on mineral element concentrations in smooth bromegrass. Agron. J. 79:774–778.

Casler, M.D., and N.J. Ehlke. 1986. Forage yield and yield component response to divergent selection of in vitro dry matter digestibility in smooth bromegrass. Crop Sci. 26:478–481.

Casler, M.D., H. Talbert, A.K. Forney, N.J. Ehlke, and J.M. Reich. 1987. Genetic variation for rate of cell wall digestion and related traits in first cut smooth bromegrass. Crop Sci. 27:935–939.

Chapman, H.D., W.H. Marchant, G.W. Burton, and W.G. Monson. 1971. Performance of steers grazing Pensacola bahia, Coastal and Coastcross 1 bermudagrass. Abstr. J. Anim. Sci. 32:374.

Chenost, M. 1966. Fibrousness of forages: Its determination and its relation to feeding value. p. 406–411. *In* A.G.G. Hill (ed.) Proc. Int. Grassl. Congr., 10th, Helsinki, Finland. 7–16 July. Finnish Grassl. Assoc., Helsinki.

Cherney, J.H., K.J. Moore, J.J. Volenec, and J.D. Axtell. 1986. Rate and extent of digestion of cell wall components of brown-midrib sorghum species. Crop Sci. 26:1055–1059.

Christensen, D.W., D.D. Stuthman, and A.W. Hovin. 1984. Associations among morphological and digestibility characters in reed canarygrass. Crop Sci. 24:675–678.

Council for Agricultural Science and Technology. 1986. Forages: Resources for the future. Rep. 108, March. CAST, Ames, IA.

Currier, C.G., R.L. Haaland, C.S. Hoveland, C.B. Elkins, and J.W. Odom. 1983. Breeding for higher magnesium content in orchardgrass (*Dactylis glomerata* L.). p. 127–129. *In* J.A. Smith and V.W. Hays (ed.) Proc. 14th Int. Grassl. Congr., Lexington, KY. 15–24 June, 1981. Westview Press, Boulder, CO.

de Ruiter, J.M., and J.C. Burns. 1987a. Digestible and indigestible cell wall carbohydrates of flaccidgrass, tall fescue, and Coastal bermudagrass. Crop Sci. 27:132–138.

de Ruiter, J.M., and J.C. Burns. 1987b. Cell wall carbohydrates of flaccidgrass plant parts. I. Neutral sugar composition of fermented residues. Crop Sci. 27:1057–1063.

de Ruiter, J.M., and J.C. Burns. 1987b. Cell wall carbohydrates of flaccidgrass plant parts. II. Neutral sugar relationships to in vitro disappearance and fiber fractions. Crop Sci. 27:1063–1068.

de Ruiter, J.M., and J.C. Burns. 1987b. Cell wall carbohydrates of flaccidgrass plant parts. III. Near infrared reflectance spectroscopy determination, rate, and extent of fermentation. Crop Sci. 27:1069–1076.

Ehlke, N.J., and M.D. Casler. 1985. Anatomical characteristics of smooth bromegrass clones selected for in vitro dry matter digestibility. Crop Sci. 25:513–517.

Ehlke, N.J., M.D. Casler, P.N. Drolsom, and J.S. Shenk. 1986. Divergent selection for in vitro dry matter digestibility in smooth bromegrass. Crop Sci. 26:1123–1126.

Fritz, J.O., R.P. Cantrell, V.L. Lechtenberg, J.D. Axtell, and J.M. Hertel. 1981. Brown midrib mutants in sudangrass and grain sorghum. Crop Sci. 21:706–709.

Gabrielsen, B.C. 1986. Evaluation of marketed cellulases for activity and capacity to degrade forage. Agron. J. 78:838–842.

Gabrielsen, B.C., K.P. Vogel, B.E. Anderson, and J.K. Ward. 1987. Soluble lignin phenolics and forage composition in switchgrass strains during grazing. p. 142. *In* Agronomy abstract. ASA, Madison, WI.

Godshalk, E.B., J.C. Burns, and D.H. Timothy. 1986. Selection for in vitro dry matter disappearance in switchgrass regrowth. Crop Sci. 26:943–947.

Gorz, H.J., F.A. Haskins, J.F. Pederson, and W.M. Ross. 1987. Combining ability effects for mineral elements in forage sorghum hybrids. Crop Sci. 27:216–219.

Grabber, J.H., and D.W. Allinson. 1985. Anatomical factors influencing in vitro digestibility of reed canarygrass. p. 14. *In* 1985 Abstract, NE Branch. ASA, Madison, WI.

Greub, L.J., M. Collins, S.K. Carlson, and M.D. Casler. 1986. Relationship of morphological characteristics to forage quality in quackgrass. Crop Sci. 26:819–822.

Hacker, J.B. 1982. Selecting and breeding better quality grasses. p. 305–326. *In* J.B. Hacker (ed.) Nutritional limits to animal production from pastures, Proc. of an Int. Symp., St. Lucia, Australia. 24–28 Aug. 1981. Commonwealth Agric. Bureaux, Farnham House, Farnham Royal, Slough, England.

Hacker, J.B., and D.J. Minson. 1981. The digestibility of plant parts. Herbage Abstr. 51:459–482.

Hanna, W.W., W.B. Monson, and G.W. Burton. 1974. Leaf surface effects on in vitro digestion and transpiration in isogenic lines of sorghum and pearl millet. Crop Sci. 14:837–838.

Haskins, F.A., H.J. Gorz, and B.E. Johnson. 1987. Seasonal variation in leaf hydrocyanic acid potential of low- and high-dhurrin sorghums. Crop Sci. 27:903–906.

Hatfield, R. 1987. Structural polysaccharides in forages and their digestibility. p. 93. *In* Agronomy abstract. ASA, Madison, WI.

Holt, E.C., and B.E. Conrad. 1986. Influence of harvest frequency and season on bermudagrass cultivar yield and forage quality. Agron. J. 78:433–436.

Holt, E.C., W.C. Ellis, and G.R. Engdahl. 1979. Forage sampling factors influencing the variability of in vitro fermentation results of grass selections. Crop Sci. 19:219–222.

Hoveland, C.S., S.P. Schmidt, C.C. King, and E.M. Clark. 1984. Association of fungal endophyte with seasonal gains of beef steers grazing tall fescue pasture. p. 382–386. *In* H. Riley and A.O. Skjelvag (ed.) The impact of climate on grass production and quality. Proc. 10th General Meet. European Grassl. Federation. The Norwegian State Agric. Res. Stn., As, Norway.

Hovin, A.W., and G.C. Marten. 1983. MN-76 low-alkaloid reed canarygrass germplasm. Crop Sci. 23:1017–1018.

Hovin, A.W., C.M. Rincker, and G.M. Wood. 1966. Breeding of nonflowering orchardgrass, *Dactylis glomerata* L. Crop Sci. 6:239–241.

Jones, D.I.H., R.J. K. Walters, and E.L. Breese. 1973. The evolution of herbage breeding programmes for improved voluntary intake and other nutritive characteristics. p. 111–120. *In* Proc. 5th General Meet., European Grassl. Federation, Uppsala, Sweden.

Jones, T.A. 1987. Compatibility of reed canarygrass with alfalfa and birdsfoot trefoil. Ph.D. thesis, Iowa State Univ., Ames. (Diss. Abstr. 47:7–8, 3172-B).

Jung, H.G. 1987. Forage lignins and their effects on fiber digestibility. p. 95. *In* Agronomy abstract. ASA, Madison, WI.

Jung, H.G., and G.C. Fahey, Jr. 1983. Nutritional implications of phenolic monomers and lignin: A review. J. Anim. Sci. 57:206–219.

Karn, J.F., and J.M. Krupinsky. 1983. Chemical composition of intermediate wheatgrass affected by foliar diseases and stem smut. Phytopathology 73:1152–1155.

Kasperbauer, M.J., D.L. Karlen, and H.R. Burton. 1987. Ploidy effects on protein, in vitro dry matter disappearance, and K/(Ca + Mg) equivalent ratio in tall fescue forage. Crop Sci. 27:1081–1082.

Lamb, J.F.S., K.P. Vogel, and P.E. Reece. 1984. Genotype and genotype × environment interaction effects on forage yield and quality of crested wheatgrasses. Crop Sci. 24:559–564.

Lane, D.W.A., and C.A. Lamp. 1980. Preliminary selection of ryegrass plants for Mg accumulating ability. p. 385–387. *In* E. Wojahn and H. Thons (ed.) Proc. 13th Int. Grassl. Congr. Vol. 1, Leipzig, East Germany. 18–27 May 1977. Akademic-Verlag, Berlin.

Marten, G.C. 1981a. Chemical, in vitro, and nylon bag procedures for evaluating forage in the USA. p. 39–55. *In* J.L. Wheeler and R.D. Mochrie (ed.) Forage evaluation: Concepts and techniques. Proc. of a Workshop, Armidale, NSW, Australia. 27–31 Oct. 1980. Am. Forage and Grassl. Counc., Belleville, PA.

Marten, G.C. 1981b. Effect of deleterious compounds on animal preference for forage and on animal performance. p. 225–235. *In* J.L. Wheeler and R.D. Mochrie (ed.) Forage evaluation: Concepts and techniques. Proc. of a Workshop, Armidale, NSW, Australia. 27–31 Oct. 1980. Am. Forage and Grassl. Counc., Belleville, PA.

Marten, G.C. 1985a. Factors influencing feeding value and effective utilization of forages for animal production. p. 89–97. *In* Proc. Int. Grassl. Congr., 15th, Kyoto, Japan. 24–31 Aug. The Natl. Grassl. Res. Inst., Nishi-narino, Japan.

Marten, G.C. 1985b. Reed canarygrass. p. 207–216. *In* M.E. Heath et al. (ed.) Forages—The science of grassland agriculture. Iowa State Univ. Press, Ames.

Marten, G.C. 1985c. Current applications of NIRS technology in forage research. p. 45–48. *In* G.C. Marten et al. (ed.) Near infrared reflectance spectroscopy (NIRS): Analysis of forage quality. USDA Agric. Handb. 643. U.S. Gov. Print. Office, Washington, DC.

Marten, G.C., and R.F Barnes. 1980. Prediction of energy digestibility of forages with in vitro rumen fermentation and fungal enzyme systems. p. 61–71. *In* W.J. Pigden et al. (ed.) Standardization of analytical methodology for feeds. Int. Devel. Res. Centre, Ottawa, Canada. IDRC-134e.

Marten, G.C., J.L. Halgerson, and D.A. Sleper. 1988. Near infrared reflectance spectroscopy evaluation of ruminal fermentation and cellulase digestion of diverse forages. Crop Sci. 28:163–167.

Marten, G.C., R.M. Jordan, and A.W. Hovin. 1976. Biological significance of reed canarygrass alkaloids and associated palatability variation to grazing sheep and cattle. Agron. J. 68:909–914.

Marten, G.C., R.M. Jordan, and A.W. Hovin. 1981. Improved lamb performance associated with breeding for alkaloid reduction in reed canarygrass. Crop Sci. 21:295–298.

Marten, G.C., J.S. Shenk, and F.E. Barton II (ed.). 1985. Near infrared reflectance spectroscopy (NIRS): Analysis of forage quality. USDA Agric. Handb. 643. U.S. Gov. Print. Office, Washington, DC.

Marum, P., and A.W. Hovin. 1979. Cell wall constituents in plant parts of reed canarygrass clones. Crop Sci. 19:280–284.

Marum, P., A.W. Hovin, and G.C. Marten. 1979a. Inheritance of three groups of indole alkaloids in reed canarygrass. Crop Sci. 19:539–544.

Marum, P., A.W. Hovin, G.C. Marten, and J.S. Shenk. 1979b. Genetic variability for cell wall constituents and associated quality traits in reed canarygrass. Crop Sci. 19:355–360.

Matches, A.G. (ed.). 1973. Antiquality components of forages. CSSA Spec. Publ. 4, CSSA, Madison, WI.

Mertens, D.R. 1982. Using neutral detergent fiber to formulate dairy rations. p. 116–126. *In* Proc. 1982 Georgia Nutr. Conf. for the Feed Industry. Univ. of Georgia, Athens.

Minson, D.J. 1980. Predicting forage intake by laboratory methods. p. 369–371. *In* E. Wojahn and H. Thons (ed.) proc. 13th Int. Grassl. Congr. Vol. 1, Leipzig, East Germany. 18–27 May 1977. Akademie-Verlag, Berlin.

Minson, D.J. 1981a. An Australian view of laboratory techniques for forage evaluation. p. 57–71. *In* J.L. Wheeler and R.D. Mochrie (ed.) Forage evaluation: Concepts and techniques. Proc. of a Workshop, Armidale, NSW, Australia. 27–31 Oct. 1980. Am. Forage and Grassl. Counc., Belleville, PA.

Minson, D.J. 1981b. Characteristics of the nutritionally ideal grazing plant. p. 543–550. *In* J.L. Wheeler and R.D. Mochrie (ed.) Forage evaluation: Concepts and techniques. Proc. of a Workshop, Armidale, NSW, Australia. 27–31 Oct. 1980. Am. Forage and Grassl. Counc., Belleville, PA.

Minson, D.J. 1982. Effects of chemical and physical composition of herbage eaten upon intake. p. 167–182. *In* J.B. Hacker (ed.) Nutritional limits to animal production from pastures. Commonwealth Agric. Bureaux, Farnham Royal, Slough, England.

Minson, D.J. 1983. Forage quality: Assessing the plant-animal complex. p. 23–27. *In* J.A. Smith and V.W. Hays (ed.) Proc. 14th Int. Grassl. Congr., Lexington, KY. 15–24 June 1981. Westview Press, Boulder, CO.

Monson, W.G., W.W. Hanna, and T.P. Gaines. 1986. Effects of rust on yield and quality of pearl millet forage. Crop Sci. 26:637–639.

Morrison, I.M. 1979. Carbohydrate chemistry and rumen digestion. Proc. Nutr. Soc. 38:269–274.

Morrison, I.M. 1980. Changes in the lignin and hemicellulose concentrations of ten varieties of temperate grasses with increasing maturity. Grass Forage Sci. 35:287–293.

Mott, G.O. 1983. Potential productivity of temperate and tropical grassland systems. p. 35–41. *In* J.A. Smith and V.W. Hays (ed.) Proc. 14th Int. Grassl. Congr., Lexington, KY. 15–24 June 1981. Westview Press, Boulder, CO.

Nasinec, J., and V. Mika. 1980. The inheritance of sodium content in the dry matter of *Dactylis glomerata* L. p. 395–397. *In* E. Wojahn and H. Thons (ed.) Proc. 13th Int. Grassl. Congr. Vol. 1, Leipzig, East Germany. 18–27 May 1977. Akademic-Verlag, Berlin.

Nelson, L.R., and F.M. Rouquette, Jr. 1983. Neutral detergent fiber and protein levels in diploid and tetraploid ryegrass forage. p. 234–236. *In* J.A. Smith and V.W. Hays (ed.) Proc. 14th Int. Grassl. Congr., Lexington, KY. 15–24 June 1981. Westview Press, Boulder, CO.

Nguyen, H.T., D.A. Sleper, and A.G. Matches. 1982. Inheritance of forage quality and its relationship to leaf tensile strength in tall fescue. Crop Sci. 22:67–72.

Obersheim, R.L., C.C. Berg, R.T. Sherwood, and K.E. Zeiders. 1987. Yield and quality of forage from orchardgrass selected for resistance to purple leaf spot. Crop Sci. 27:673–676.

Oram, R.N., H.E. Schroeder, and R.A. Culvenor. 1985. Domestication of *Phalaris aquatica* as a pasture grass. p. 220–221. *In* Proc. Int. Grassl. Congr., 15th, Kyoto, Japan. 24–31 Aug. The Natl. Grassl. Res. Inst., Nishi-narino, Japan.

Paul, C., U.V. Borstel, and E. Zimmer. 1985. Using grinding energy measurements for predicting crude fibre in first growths of permanent grassland. p. 955–956. *In* Proc. Int. Grassl. Congr., 15th, Kyoto, Japan. 24–31 Aug. The Natl. Grassl. Res. Inst., Nishi-narino, Japan.

Pedersen, J.F., H.J. Gorz, F.A. Haskins, and W.M. Ross. 1982. Variability for quality and agronomic traits in forage sorghum hybrids. Crop Sci. 22:853–856.

Reed, K.F.M. 1981. A review of legume-based vs. nitrogen-fertilized pasture systems for sheep and beef cattle. p. 401–417. *In* J.L. Wheeler and R.D. Mochrie (ed.) Forage evaluation: Concepts and techniques. Am. Forage and Grassl. Counc. and CSIRO, Australia.

Reeder, L.R., Jr., H.T. Nguyen, D.A. Sleper, and J.R. Brown. 1986. Genetic variability of mineral concentrations in tall fescue grown under controlled conditions. Crop Sci. 26:514–518.

Reich, J.M., and M.D. Casler. 1985. Genetic variation for response to advancing maturity of smooth bromegrass forage quality traits. Crop Sci. 25:641–646.

Reid, R.L., and W.C. Templeton, Jr. 1984. Different combinations of grasses and legumes for growing lambs. p. 161. *In* Agronomy abstract. ASA, Madison, WI.

Saiga, S. 1983. Selection effect for digestibility of summer regrowth in orchardgrass. p. 143–146. *In* J.A. Smith and V.W. Hays (ed.) Proc. 14th Int. Grassl. Congr., Lexington, KY. 15–24 June 1981. Westview Press, Boulder, CO.

Shenk, J.S. 1977. The role of plant breeding in improving the nutritive value of forages. J. Dairy Sci. 60:300–305.

Short, K.E., and I.T. Carlson. 1987. Genetic variation and response to selection for legume-compatibility traits in orchardgrass. p. 79. *In* Agronomy abstract. ASA, Madison, WI.

Simon, U., and P. Daniel. 1983. Effect of experimental methods on results of voluntary intake experiments with grass cultivars. p. 489–491. *In* J.A. Smith and V.W. Hays (ed.) Proc. 14th Int. Grassl. Congr., Lexington, KY. 15–24 June 1981. Westview Press, Boulder, CO.

Sleper, D.A. 1985. Breeding tall fescue. p. 313–342. *In* Plant breeding reviews. Vol. 3. AVI Publ. Co., Westport, CT.

Sleper, D.A., G.B. Garner, K.H. Asay, R. Boland, and E.E. Pickett. 1977. Breeding for Mg, Ca, K, and P content in tall fescue. Crop Sci. 17:433–438.

Sleper, D.A., and P.G. Roughan. 1984. Histology of several cool-season forage grasses digested by cellulase. N.Z. J. Agric. Res. 27:161–166.

Soh, A.C., R.V. Frakes, D.O. Chilcote, and D.A. Sleper. 1984. Genetic variation in acid detergent fiber, neutral detergent fiber, hemicellulose, crude protein, and their relationship with in vitro dry matter digestibility in tall fescue. Crop Sci. 24:721–727.

Stratton, S.D., D.A. Sleper, and A.G. Matches. 1979. Variation and interrelationships of in vitro dry matter disappearance and fiber content in orchardgrass herbage. Crop Sci. 19:329–333.

Sullivan, J.T., E.S. Titlar, and T.V. Hershberger. 1962. Chemical composition of nonheading orchardgrass plants. J. Anim. Sci. 21:138.

Tarumoto, I., and Y. Masaoka. 1983. A simple method of evaluating forage digestibility and its application to sorghum breeding. p. 146–149. *In* J.A. Smith and V.W. Hays (ed.) Proc. 14th Int. Grassl. Congr., Lexington, KY. 15–24 June 1981. Westview Press, Boulder, CO.

Twidwell, E.K., K.D. Johnson, C.E. Bracker, J.A. Patterson, and J.H. Cherney. 1987. Use of image analysis to measure plant anatomical degradation. p. 146. *In* Agronomy abstract. ASA, Madison, WI.

Ugherughe, P.O., P.N. Drolsom, and J.R. Davis. 1980. Influence of planting pattern on estimated digestibility of smooth bromegrass. Crop Sci. 20:695–699.

Ulyatt, M.J., and J.L. Black. 1984. Plant breeding and feeding value of pastures and forages. CSIRO-DSIR Workshop Summary. DSIR, Palmerston North, New Zealand.

van Bogaert, G. 1977. Scope for improving the yield of grasses and legumes by breeding and selection. p. 29–35. *In* Proc. Int. Meet. on Anim. Production from Temperate Grassl. Dublin, Ireland. June. Irish Grassl. and Animal Production Assoc., Agric. Inst., Belclare, Tuam, Co. Galway, Ireland.

Van Soest, P.J., and D.R. Mertens. 1977. Analytical parameters as guides to forage quality. p. 50–61. *In* Proc. Int. Meet. on Animal Production from Temperate Grassl. Dublin, Ireland. June. Irish Grassl. and Animal Production Assoc., Agric. Inst., Belclare, Tuam, Co. Galway, Ireland.

Vincent, J.F.V. 1983. The influence of water content on the stiffness and fracture properties of grass leaves. Grass Forage Sci. 38:107–114.

Vogel, K.P., R. Britton, H.J. Gorz, and F.A. Haskins. 1984. *In vitro* and *in vivo* analyses of hays of switchgrass strains selected for high and low IVDMD. Crop Sci. 24:977–980.

Vogel, K.P., F.A. Haskins, and H.J. Gorz. 1981. Divergent selection for in vitro dry matter digestibility in switchgrass. Crop Sci. 21:39–41.

Vogel, K.P., P.E. Reece, and J.F.S. Lamb. 1986. Genotype and genotype × environment interaction effects for forage yield and quality of intermediate wheatgrass. Crop Sci. 26:653–658.

Voigt, P.W. 1975. Improving palatability of range plants. p. 23–49. *In* R.S. Campbell and C.H. Herbel (ed.) Improved range plants. Symp. Ser. 1. Am. Soc. Range Manage., Denver, CO.

Voigt, P.W., W.R. Kneebone, E.H. McIlvain, M.C. Shoop, and J.E. Webster. 1970. Palatability, chemical composition, and animal gain from selections of weeping lovegrass, *Eragrostis curvula* (Schrad.) Nees. Agron. J. 62:673–676.

Welch, J.G. 1986. Physical parameters of fiber affecting passage from the rumen. J. Dairy Sci. 69:2750–2754.

Williams, M.J., P.A. Backman, E.M. Clark, and J.F. White. 1984. Seed treatments for control of the tall fescue endophyte *Acremonium coenophialum*. Plant Dis. 68:49–52.

Wilson, J.R. 1985. An interdisciplinary approach for increasing yield and improving quality of forages. p. 49–54. *In* Proc. Int. Grassl. Congr., 15th, Kyoto, Japan. 24-31 Aug. The Natl. Grassl. Res. Inst., Nishi-narino, Japan.

Wilson, J.R., R.H. Brown, and W.R. Windham. 1983. Influence of leaf anatomy on the dry matter digestibility of C_3, C_4, andC_3/C_4 intermediate types of *Panicum* species. Crop Sci. 23:141–146.

7 Breeding Grasses for the Future[1]

K. P. Vogel, H. J. Gorz, and F. A. Haskins

USDA-ARS
University of Nebraska
Lincoln, Nebraska

ABSTRACT

Plant breeding, including grass breeding, involves taking a raw product, plant germplasm, and improving or adding value to that germplasm by manipulating its genetic composition. The value added to the germplasm has a cost. It usually costs in excess of $100 000/yr to maintain a viable, ongoing grass breeding program. The output of a grass breeding program, i.e., the released cultivars and germplasm, should have an economic value in excess of the cost of the breeding program. Grass breeding programs have produced products such as 'Coastal' bermudagrass [*Cynodon dactylon* (L.) Pers.] where the economic value has greatly exceeded the input cost. Grass breeders have the opportunity to make additional major contributions to the welfare and benefit of future generations of humanity if research goals are carefully delineated and innovative, cost-effective breeding methods are used.

Plant breeders take a raw product, plant germplasm, and by genetic manipulations produce products such as cultivars that are of increased value to humanity. Hence, plant breeding can be described as human-directed evolution. The process is analogous to manufacturing in which raw materials are transformed into items that are of increased value by the manufacturing process. In each instance, value is added to the raw material by the breeding or manufacturing process. This improved value fulfills specific needs of humanity. Cultivars and other products of plant breeding must fulfill specific needs of society if they are to have any value.

The value that is added to germplasm in the breeding process occurs in incremental steps. A superior grass genotype or plant growing on the plains of Africa has value only to the owner of the cow (*Bos taurus*) that may graze it once or twice a year or to a wild herbivore. Although it has intrinsic value, its economic value is limited. If this plant is collected, increased, evaluated, used as a source of desirable genes in a breeding program, production tested, increased and released as a cultivar that is planted on millions of acres, its

[1] Joint contribution of USDA-ARS and the Univ. of Nebraska, Lincoln, NE 68583.

value is greatly multiplied. This procedure was basically used in the development of 'Coastal' bermudagrass [*Cynodon dactylon* (L.) Pers.] that is planted on millions of acres in the southern USA and that has added billions of dollars to the economy of this and other countries (see Chapter 3 in this book). Similar examples could be given for many other grasses. Each step in this breeding procedure (Fig. 7–1) adds value to the germplasm. As in the example above, uncollected germplasm has little economic value; it has value to humanity only after it has been collected and its attributes are known.

Each step in the development of an improved cultivar adds value, but the value added has a cost just as in manufacturing. Each step requires trained personnel, equipment, and facilities. In many grass breeding programs, the breeders may be involved in all aspects of cultivar development that are illustrated in Fig. 7–1. Although each step in the breeding procedure adds benefits and has a cost, the value added does not directly benefit humanity nor is there any return on the "investment" until an improved cultivar is released and used.

GERMPLASM
↓
$ ↓
↓
↓
COLLECTION
↓
$ ↓
↓
↓
EVALUATION
↓
$ ↓
↓
↓
ENHANCEMENT
↓
$ ↓
↓
↓
BREEDING
↓
$ ↓
↓
↓
PRODUCTION TESTING
↓
$ ↓
↓
↓
CULTIVAR INCREASE AND RELEASE
↓
$ ↓
↓
↓
UTILIZATION

Fig. 7–1. Steps in the development and release of a forage or turf cultivar.

Grass breeders have developed cultivars that have greatly benefitted humanity and for which the economic value of the cultivar has greatly exceeded the development cost (see Chapters 2, 3, 4, and 5 in this book). The future of grass breeding depends upon the capability of grass breeders to continue to develop products or cultivars whose value greatly exceeds the investment cost. In general, it costs a minimum of $100 000/yr to maintain an ongoing grass breeding program. Since most turf and forage grasses are perennials that require extended periods for breeding and evaluation, a minimum period of 10 yr is usually required to develop a new cultivar. Thus, a new grass cultivar may represent an investment in excess of one million dollars. Its value to society and the public institution or private firm that developed it must exceed this cost. We recognize that public breeding programs also involve the training of professionals and the publication of research results that add to the information pool, but the economic impact of these products is difficult to assess. In this report, we will identify breeding objectives and appropriate breeding procedures for turf and forage grasses that in our opinion can have substantial economic impact. We have not attempted a comprehensive review of the literature, but rather have chosen specific papers that we have used to illustrate a point. We will also discuss selection, evaluation, and breeding methods that in our opinion have the most potential for making significant gains by breeding.

FUTURE BREEDING OBJECTIVES WITH POTENTIAL ECONOMIC IMPACT

Turf and forage grass breeders attempt to improve one or more plant attributes in their breeding programs. The principal attributes and the potential benefits that can be obtained by improving these attributes are as follows.

Establishment

Rapid and reliable establishment is critical to the economic culture of turf and forage grasses. Seedling establishment can be improved by either improving the establishment capability of the plants by breeding or by modifying the environment with cultural practices. The primary causes of poor establishment are the related factors of moisture stress and weed competition. Breeding for improved seedling vigor can result in seedlings that develop rapidly and that are effective competitors with weeds for available moisture. Factors that affect seedling vigor are seed size, seed quality, germination rate, emergence rate, relative growth rates, and other physiological processes (McKell, 1972). In virtually all the studies that have been done to date, seed size or weight has been an important component of seedling establishment capability (Asay & Johnson, 1987; Voight et al., 1987). For example, in sand bluestem [*Andropogon gerardii* var. *Paucipilus* (Nash) Fern.], 50% of the genetic variability for seedling weight 8 wk after emergence was due to differences in seed size, while the remainder of the genetic variability was due to other factors (Glewen & Vogel, 1984). Many physiological processes are involved in seedling establishment. Seedling weight at a fixed time following seeding is a simple method of quantifying the end result of these processes and making selections (Asay & Johnson, 1987; Voight et al., 1987). Selections for seedling size or weight are inexpensive and effective methods of improving vigor. In most forage and turfgrasses, improvements in establishment capability by breeding for seedling vigor will require long-term breeding efforts. Most forage and turfgrasses are perennials and their main competitors at establishment are annual weeds, many of which excel in seedling vigor.

Breeding for mechanisms that permit effective weed control can be an effective method of improving establishment capability. Atrazine [2-chloro-4-ethylamino-6-isopropylamino-1,3,5-triazine] can be used as a preemergence herbicide for the establishment of big bluestem [*Andropogon gerardii* Vitman] and switchgrass (*Panicum virgatum* L.) (Martin et al., 1982). Preemergence atrazine applications can reduce the seeding rate needed for the establishment of satisfactory stands of these grasses with a net savings in cost of seed of $35 to $148/ha (Table 7–1), and can result in increased forage production in the establishment year with a net value of more than $220/ha (Table 7–2). Big bluestem and switchgrass are the only two forage grasses

Table 7–1. Seeding rates and costs for establishing big bluestem and switchgrass with preemergence atrazine application (2.2 kg ha^{-1}).†

	Switchgrass			Big bluestem		
Seeding rate	Stand	Seeding rate	Cost‡	Stand	Seeding rate	Cost
seed m^{-2}	%	kg ha^{-1}	\$ ha^{-1}	%	kg ha^{-1}	\$ ha^{-1}
215	66	3.4	75	69	6.7	147
325	71	5.0	110	65	10.1	220
430	70	6.7	147	78	13.4	295
F value	NS			NS		

† Data except for costs are from Vogel (1987).
‡ Seed costs estimated at $22 kg^{-1} ($10 lb^{-1}) for both grasses.

Table 7–2. Effect of preemergence atrazine applications on yield and net return of big bluestem and switchgrass at Mead, NE. Forage yields are from Martin et al. (1982).

		Yield†		Net return‡	
Grass	Atrazine	Yr 1	Yr 2	Yr 1	Yr 2
	kg ha^{-1}	Mg ha^{-1}		\$ ha^{-1}	
Big bluestem	0	0.8	7.1	40	355
	2.2	7.2	9.1	340	455
Switchgrass	0	0	9.0	0	450
	2.2	5.4	11.3	250	565

† Years 1 and 2 are the year of establishment and the year following establishment, respectively.
‡ Based on a hay value of $50 Mg^{-1} and an atrazine cost of $20 ha^{-1}.

for which a preemergence herbicide is labeled for use. The atrazine tolerance of these grasses has a potential value in excess of 100 million dollars because of the reduced seeding costs and increased establishment-year forage yields that can be achieved if atrazine is used as a preemergence herbicide. Breeding for herbicide tolerance in other grasses by using either conventional or "genetic engineering" techniques could have a similar impact. Breeding for herbicide tolerance may result in greater improvement in establishment than breeding for seedling vigor or seed size.

Persistence

Breeding for persistence is a worthwhile objective since the annual cost of establishment and the subsequent loss of production and use equals the establishment costs divided by N, where N is the number of years the stand persists. Grass breeders traditionally have selected for persistence by first making selections among species for adaptation to the target environment and cultural conditions (Hanson & Carnahan, 1956). Because of the large number of grasses that were available, it was usually easier to select at the species level for a particular ecological niche than to attempt to change the adaptability of a species by breeding. Breeders found adapted germplasm in either native species or in introduced species from areas that were approximate climatic analogues of the target environment. They have subsequently conducted breeding work to improve disease and insect resistance because of the effect these factors have on persistence. Increased knowledge of the physiological processes that affect persistence has enabled breeders to devise screening procedures that can aid in selecting for persistence. However, there are no short cuts in breeding for persistence; extended testing in realistic stress situations is required.

Disease and Insect Resistance

Diseases affect turf and forage grass persistence, quality, yield, and utilization. Turfgrass diseases directly affect every homeowner in the USA who has a lawn. Control of turf diseases is a multi-million dollar a year business that helps to support a growing proliferation of lawn and garden stores and professional turf maintenance services. Diseases of forage grasses also affect everyone, except strict vegetarians, by their effect on forage yields and quality which directly affect animal production of meat and milk. Grass breeders have made significant improvements in the disease tolerance and resistance of turf and forage grasses. Virtually every improved cultivar on the market today is superior in disease resistance to common strains or earlier cultivars. Additional genetic gains in disease resistance can be made for turf and forage grasses. Genetic sources of resistance have been reported for almost every disease of economically important cool-season grasses (Braverman, 1986). A similar situation probably exists in warm-season grasses. Breeding for disease resistance should be an integral part of every grass breeding program. In the perennial grass breeding program at Lincoln, NE, we continually apply selection pressure for disease resistance, generation after generation. More structured programs may be needed for severe disease problems, in which case a plant pathologist should be an integral part of the breeding team. When new, disease-resistant cultivars are developed, breeders need to document the economic benefits to homeowners or forage producers in terms that a layman can readily understand.

Insects also have a major impact on turf and forage grasses. In Lincoln, homeowners have to spend from $30 to $50 per year for insect control if they want to have an acceptable bluegrass (*Poa pratensis* L.) lawn. Damage by other insects such as the *Labops* spp. on wheatgrasses also can produce tremendous economic losses (Campbell et al., 1984). Genetic sources for insect tolerance or resistance have been identified for important turf and forage insect pests (Stimman & Taliaferro, 1969; Asay et al., 1983; Campbell et al., 1984). Recent reports (Funk et al., 1983; Johnson et al., 1985; Clay et al., 1985) demonstrate that fescue (*Festuca* spp.) and ryegrass (*Lolium* spp.) endophytes (*Acremonium* spp.) confer broad-spectrum insect resistance to grasses they infest. These endophytes will reduce the economic losses to turf insects by billions of dollars per year once they have been incorporated into fescue, ryegrass, and possibly other grasses that are widely used for turf. Breeding for insect resistance can have significant economic benefit and the use of resistant cultivars is ecologically more desirable than using insecticides for insect control.

Seed Yield

Considerable effort has been and is being expended to improve the seed yield of forage and turf grasses. Except for specific instances where poor seed yields are preventing the use of an otherwise valuable grass, we do not believe that this effort is warranted. Seed yields for grasses are lower than for grain crops, but the number of hectares of forage crops for use as pastures or harvested forage that can be seeded from each seed production hectare is often greater for grasses than for grain crops (Table 7–3). At present prices, the per hectare value of the seed produced on most grass seed fields is greater than the per hectare value of seed produced from certified grain crops, and

Table 7–3. Comparison of wheat, soybean, and grass certified seed yields and production field or pasture establishment costs.†

	Seed field		Production field or pasture			Seed yield/ seeding rate ratio
Crop	Seed yield	Value	Seeding rate	Cost		
	kg ha^{-1}	\$ ha^{-1}	kg ha^{-1}	\$ ha^{-1}	\$ ha^{-1} yr^{-1}	
Tall fescue	600	1320	12	53	5	50
Switchgrass	400	7040	3.5	77	7	114
Wheat	2700	520	67	13	13	40
Soybean	2700	832	67	21	21	40

† Average yields and current prices in principal areas of production are listed. Grass stands were assumed to persist for 10 yr.

the cost of establishing pastures or other forage crops is less than that of grain fields when amortized over years (Table 7–3). Grass seed costs are currently high because of the Conservation Reserve Program, but even if they were half of their present price, grass seed production would still be economical for growers. Since seed is not the principal product of these grasses and since seed yields of most grasses are adequate in general, it does not seem reasonable to routinely breed for improved seed yields. There are instances, however, in which breeding for seed yield can improve the use of a grass; for example, the development of cultivars with reduced seed shattering in reed canarygrass [*Phalaris arundinaceae* L.] (R.R. Kalton, 1987, personal communication), and the discovery of a mutant in eastern gamagrass [*Tripsacum dactyloides* (L.) L.] that increases seed yield 20- to 25-fold (Dewald & Dayton, 1985). In each of these examples, a specific problem that was limiting seed yield was solved and was worth the breeding effort expended.

Forage Yield

Improving forage yield has always been one of the principal objectives of grass breeders. Recent reports document that forage yields can be significantly improved by breeding with substantial economic benefits (Table 7–4). Burton (1982, 1985) improved forage yield of bahiagrass (*Paspalum notatum* Fluegge) by direct selection for yield using Restricted Recurrent Phenotypic Selection. Nelson et al. (1985) improved forage yield of tall fescue [*Festuca arundinaceae* Schreb.] by selection for leaf area expansion rate. These breeding efforts were successful because the breeders either selected directly for yield or for a trait that was correlated with yield, and they used recurrent selection methods that effectively exploited the additive genetic variability for the selected traits within the species. It should be possible to improve the yield of most forage grasses by using well-designed recurrent selection methods. Breeding for yield remains a valid research objective.

Table 7–4. Economic gains achieved by breeding for forage yield in bahiagrass and tall fescue.†

Population	Forage yield		Gain/cycle	
	kg ha^{-1}	\$ ha^{-1}	kg ha^{-1}	\$ ha^{-1}
Bahiagrass (Burton, 1985)				
Pensacola commercial	5273	264		
Pensacola RRPS Cycle 9	9241	462	441	22
Tall fescue‡ (Nelson et al., 1985)				
Leaf area expansion C0	5130	256		
High leaf area expansion C4	6274	314	286	14

† Yields are from references indicated. Hay value = \$50 Mg^{-1}.
‡ Mt. Vernon data.

Forage Quality

Forage quality can be improved by breeding for enhanced positive quality factors such as digestibility or for reduced negative factors such as alkaloids (Burton, 1981; see Chapter 6 in this book). Significant gains have been made in improving digestibility that have resulted in improved animal performance (Anderson et al., 1988; Chapman et al., 1972). Reduction in the levels of undesirable alkaloids either by direct selection (Marten et al., 1976) or by eliminating endophytic fungi that are associated with undesirable alkaloids from the forage (Hoveland et al., 1983) has also resulted in significantly improved animal performance. Other examples could be given for both positive and negative quality factors. It should be possible to improve the forage quality of virtually every grass by using appropriate breeding procedures since genetic variability has been reported for both positive and negative quality factors in virtually every grass that has been studied (Burton, 1981). In addition, it should be possible to simultaneously improve both forage yield and quality since the correlations between these traits in most studies are either low or nonsignificant. In those instances when correlations have been negative, the correlation coefficients have been low. If a breeder must make a choice between breeding for yield or quality, he/she should select quality. Increased quality results in increased net return to a livestock producer and does not require any additional investment. Increased yield can increase net return but the producer must buy or raise additional livestock to use the additional forage.

Turf Quality

Turf quality is often an aesthetic criterion that is difficult to quantify and weigh in economic terms in a breeding program. It can be rated or ranked, but the rating is dependent upon the personal preference and skill of the person doing the ranking. Turf quality factors such as color, leaf size, leaf texture, tiller density, wear tolerance, absence of disease and insect damage are all under genetic control. Turf breeders have improved turf quality and we are confident that they will continue to do so since quality greatly influences acceptability.

Turf Maintenance Costs

Turf requires mowing, fertilization, and weed, disease, and insect control, and in many areas of the country, irrigation. All of these practices contribute to turf maintenance costs. Cockerham and Gilbeault (1985) have reported that for a hypothetical city of 170 000 located in a major urban area of the USA, there are 2500 ha of turf of which about 1400 ha are accounted for by the lawns of approximately 45 000 homes. Homeowners will each spend more than $200 a year to maintain their lawns (Cockerham & Gilbeault, 1985). The remaining turf is for apartments, parks, churches, golf courses, cemeteries, businesses, and industrial sites. Development of turfgrasses with disease and insect resistance, slow growth rate that reduces the number of mowings, and reduced fertilizer requirements can reduce maintenance costs. If improved turf cultivars were developed that could reduce maintenance costs by 25%, the savings to home-owners alone in this hypothetical city would be in excess of 2.2 million dollars a year. A city of this size, such as Lincoln, NE could probably afford to have its own turf breeding program because of the potential savings such a program could provide its citizens.

In the western half of the USA where it is necessary to irrigate lawns, improving the water-use efficiency of turfgrasses can also greatly reduce maintenance costs and also the cost of maintaining extensive water systems. The city of Thornton, CO (population 57 000), a suburb of Denver, has spent 52 million dollars to buy farms and their water rights from an area near Fort Collins with the intention of piping this water 80 km south to Thornton (Flanery, 1987). The city of Denver and its suburbs are proposing to build a $500 million dam on the South Platte River to meet the area's future water needs (Flanery, 1987). There is substantial genetic variability among turfgrasses for water-use efficiency as measured by evaportranspiration rates (Beard, 1985; Shearman, 1986). It is reasonable to assume that developing grasses that require less water to maintain a functional and attractive turf could save cities and their citizens substantial amounts of money, particularly in the arid and semiarid western states. These states and even individual cities should consider funding or increasing the funding of turfgrass breeding programs.

IMPROVED METHODS FOR SELECTION AND EVALUATION

Selection is the component of the breeding process that usually determines the success or failure of a program. The other major component of the process, mating the selected plants, usually is done in a routine manner. Selection of the plants to be mated is the critical component of the breeding process.

Selection for Physiological Traits

Breeders can improve a quantitatively inherited trait such as yield by either direct selection for the trait or by indirect selection for component physiological processes. Plant physiologists have identified and described physiological processes such as photosynthetic rate, light interception ability, respiration, photosynthate partitioning, evaportranspiration rate, and others that determine traits such as yield. Instrumentation has been developed that enables breeders to measure and select directly for these physiological processes. In general, breeding for physiological processes is expensive and time consuming, and it has not been as successful as direct selection for the trait itself (Asay & Johnson, 1983; Kube et al., 1989). Breeding for physiological traits has not been as successful as direct selection because usually only one or two physiological parameters have been measured and used to make selections while numerous processes are involved in a complex trait such as yield. The lack of success can also be explained by examining equations for predicting gain from selection (Falconer, 1981).

$$G_x = i\ h_x\ \sigma_{ax} \qquad [1]$$

$$CG_x = i\ h_y\ r_{xy}\ \sigma_{ax} \qquad [2]$$

$$CG_x = G_x \text{ if } [(h_y r_{xy})/h_x] > 1 \qquad [3]$$

where G is the expected gain from selection for traits x or y (subscripts); CG_x is a correlated response in x due to selection for y; i is the selection intensity; h is the square root of heritability, σ_a is the square root of additive genetic variability for trait x or y (subscripts) and r_{xy} is the genetic correlation between traits x and y. Equation [1] gives the expected gain for direct selection for a trait such as yield (x) while Eq. [2] gives the expected gain for a trait such as yield (x) when selecting for another trait such as photosynthetic rate (y). Comparing Eq. [1] and [2] as a ratio (Eq. [3]) indicates that indirect selection can be as effective as direct selection only if h_y for the physiological trait is at least 25% larger than that of h_x and the genetic correlation is 0.8 or larger. This usually does not occur and hence direct selection is more efficient. Physiological studies have been extremely helpful, however, by providing information that has enabled breeders to modify or alter the selection environment so as to provide maximum differentiation between genotypes. A practical example is the use of irrigation gradients to select for drought tolerance.

Selection for Plant Composition

The development of instruments and associated software and procedures for such methodology as near infrared reflectance spectroscopy (NIRS), high performance liquid chromatography (HPLC), gas chromatography, ion chromatography, and other procedures has greatly expanded the capability of both turf and forage grass breeders to select for specific plant composition and constituents. The objectives of turf and forage breeders are often diametrically opposite. For example, turf breeders may want high fiber content because it improves wear tolerance (Shearman & Beard, 1975) while forage breeders want low fiber content because it is associated with improved digestibility. It is probably true that regardless of the specific plant constituent a breeder wants to measure and select for or against, analytical equipment is now available for the necessary analyses. These analyses, however, are not cheap because the equipment is expensive to purchase, maintain, and operate.

Breeders usually need to be able to rank genotypes for a particular plant constituent, and precise analytical values are usually unnecessary if the rankings are adequate. It may be possible to reduce the analytical costs for breeder samples by modifying the procedures so that they adequately rank the sam-

Table 7-5. Performance of beef yearlings grazing switchgrass at Mead, NE in 1982, 1983, and 1985 (3-yr means).†

Strain	IVDMD	Available forage	Animal gain	Gross return‡
	%	kg ha $^{-1}$		\$ ha $^{-1}$
Trailblazer	58.0	3420	351	463
Pathfinder	56.2	3380	284	374
Low-IVDMD	55.5	3160	299	394

† IVDMD%, available forage, and animal gain values are from Anderson et al. (1988).
‡ Gain was valued at \$1.32/kg (\$0.60/lb).

ples but with a lower precision. If a breeder has limited funds for analyses he would probably make more gains by analyzing more samples at a lower degree of precision than a few samples with a high degree of precision. For example, with NIRS analyses for a trait such as digestibility, it has been recommended that for the most precision, calibration samples should be developed each year for each selection nursery. In our grass breeding program at Lincoln, we have developed general NIRS prediction equations for digestibility using samples collected from a broad array of genotypes during the growing season for 2 yr. The use of a general equation for each species will enable us to save the funds and time required to develop new calibrations each year for each nursery. We are accepting a lower level of precision, but since the standard errors of our control samples are less than they were with conventional IVDMD analyses, we are comfortable with the results.

A forage cultivar has value only when it is used. Laboratory analyses can be used to breed cultivars with improved quality, but the results of laboratory analyses will not convince a farmer or rancher to plant a cultivar with improved digestibility. Actual animal performance data are needed that quantify the genetic gains in economic terms. In small plots, the new switchgrass cv. Trailblazer had forage yields similar to the cv. Pathfinder, but it differed in digestibility by 3 to 4 percentage units (Vogel et al., 1981, 1984). In replicated pasture trials with yearling steers, cattle grazing Trailblazer had higher total gains/ha which resulted in \$89/ha (\$35/acre) greater net return for the Trailblazer pastures than for the Pathfinder pastures (Table 7-5). Foundation and certified seed of Trailblazer has sold out every year since its release and it is obvious that the pasture and not the small plot data are responsible for its demand. It cost approximately \$12 000 for materials and labor to develop the set of 12 permanent 0.4-ha pastures that were used in the Trailblazer grazing study, which includes the cost of buried water lines and water tanks on concrete pads. These pastures will be usable for other studies since they have an expected life span of 30 yr. Pastures for evaluation of breeding material are probably the single best investment of research dollars that a grass breeding program can make.

Selection for Nebulous Attributes

Traits such as tolerance to grazing and wear tolerance in turf are difficult to define, evaluate, and quantify. The only way that traits such as tolerance to grazing can be evaluated is to conduct animal trails of such duration and intensity that the trait is expressed. Turf traits such as wear tolerance also eventually have to be evaluated in realistic situations. Breeders who will make progress in improving these traits will be those who are willing to expend the effort to collect the required "real world" data.

Multiple Trait Selection

Direct selection for a single trait usually will result in the maximum gain from selection. Multiple trait selection adds to the challenge because the desired traits may have low or negative genetic correlations with one another. In general, grass breeders have made limited use of formal selection indexes based on quantitative genetic theory but have instead relied on informal subjective indexes. We believe that the use of selection indexes based on quantitative theory in which traits are weighted by realistic economic values will result in improved breeding efficiency. Considerable work needs to be done in determining economic values of traits. For example, at the present time we do not know what a unit of forage yield is worth in economic terms in comparison to a unit of digestibility for any forage grass.

CONVENTIONAL BREEDING PROCEDURES

The breeding system that a breeder uses determines the rate of gain from breeding, its cost, and the potential gain that can be made. Conventional grass breeding systems use additive and nonadditive genetic variability in plants with both sexual and apomictic reproductive mechanisms to make genetic gains.

Additive Genetic Variability

Almost all forage and turf grasses are cross-pollinated by wind. They have small florets that are difficult to emasculate, and effective mechanisms for producing hybrids such as cytoplasmic male-sterility (cms) have not been developed for most of these grasses. Thus, breeders are largely limited to procedures that use additive genetic variability and that do not require any emasculation. Fortunately, there is substantial additive genetic variability for most traits in grasses, and breeding methods that do not require emasculation are some of the most efficient that are available. The expected gain from selection that can be made by using the breeding procedures or schemes that have been developed to date are described by Empig et al. (1972), Nguyen and Sleper (1983), and Hallauer and Miranda (1981). In grass breeding programs, most of the available breeding procedures can result in breeding progress if the following guidelines are followed.

1. A productive population that possesses substantial genetic variability for the desired traits is used as the base population.
2. An adequate effective population size is maintained. The rate of inbreeding for wind-pollinated genotypes in an isolation is 1/2N where N is

the number of selected genotypes being polycrossed. If N = 100, the rate of inbreeding is 0.5% per cycle which is negligible. We consider effective population sizes of 50 (rate of inbreeding of 1% per cycle) to be the minimum population size to use in a long-term program. The number of selection units, i.e., families or plants, needed in the selection nursery is N divided by the selection percentage expressed as a decimal fraction.

3. Recurrent selection procedures are used.
4. The selected traits are quantified with reasonable precision and environmental variation is adequately controlled.

The procedure that will give the best gains will be the one that uses the most additive genetic variability per cycle or per year. Restricted Recurrent Phenotypic Selection (Burton, 1974) has the potential to make the most gains per year of any breeding procedure if the trait can be adequately measured on an individual plant basis. However, the most efficient breeding procedure probably has not yet been developed.

Nonadditive Genetic Variability

Grass breeders in general have not capitalized on the nonadditive genetic variability that exists in forage grasses even though substantial heterosis for traits such as yield exists in many grasses. The inability to effectively emasculate large numbers of plants in seed production fields has limited grass breeders' ability to develop hybrids for commercial use. One breeder, Dr. Glenn Burton, has successfully produced hybrid cultivars by using a variety of techniques as summarized in a recent review (Burton, 1986). These techniques and their possible application to other grasses are as follows.

1. First-generation chance hybrids. Four inbred lines of pearl millet [*Pennisetum glaucum* (L.) R. Br.] that flowered at the same time were bulked and used to plant seed production fields. The seed harvested from the seed field contained 75% hybrid seed of the six possible hybrids. Plots planted with this seed yielded as well as plots seeded with a mixture of the six controlled crosses because the more vigorous hybrids crowded out the less vigorous selfs and sibs (Burton, 1948). 'Gahi 1' pearl millet was released and used as a first generation hybrid until it was replaced by superior hybrids developed by using cms. This same procedure could be used for many other grasses. Since many grasses have high levels of self-incompatibility, lines based on sibs or families or even populations could be used in lieu of inbred lines.
2. Self-incompatibility hybrids. Many perennial grasses contain plants that are self-incompatible but that are cross-compatible with each other. If two plants are identified that produce superior F_1 hybrids, then the two plants can be vegetatively increased and transplanted into seed production fields. All the seed harvested from the field would be F_1-hybrid seed assuming that proper isolation requirements were maintained. These seed fields could be maintained for many years for perennial grasses. Two bahiagrass hybrids were developed based on this procedure but were not successful because of the labor and cost of establishing seed production fields. Two recent developments should improve the economics of developing self-

incompatibility hybrids. Tissue culture techniques can improve the process of vegetatively increasing individual plants, and large scale increases of single plants are possible. Transplanting procedures have been mechanized to the extent that large commercial sugarbeet [*Beta vulgaris* L.) fields are now being established with transplanted seedlings. Because of these developments, we believe that self-incompatibility hybrids of many perennial grasses could be commercially feasible.

3. Cytoplasmic male-sterile hybrids. Cytoplasmic male-sterility has been used to develop hybrids of many crops. Many desirable grasses are polyploids and in addition are self-incompatible which makes identifying and transferring maintainer and restorer genes into desirable germplasm difficult. Considerable effort by both public and private breeders has been expended to develop cms wheat (*Triticum aestivum* L.) hybrids with limited commercial success. It is doubtful that breeders of polyploid grasses could be successful in developing cms hybrids since they will have fewer resources. Cytoplasmic male sterility may be a useful breeding procedure to produce hybrids for diploid grasses, however.

4. Apomictic hybrids. Many grasses including bahiagrass, buffelgrass (*Cenchrus ciliaris* L.) and Kentucky bluegrass (*Poa pratensis* L.) produce seed by apomictic mechanisms. Basic genetic studies in these grasses have provided breeders with information that allows them to make sexual crosses that produce F_1 apomictic progeny (Hanna & Bashaw, 1987). Superior plants can be identified and once identified can be multiplied by direct seed production without any loss of vigor or change in genotype. Apomictic hybrids such as Coastal bermudagrass can also be vegetatively propogated for commercial use. Apomictic mechanisms are the most economical way to produce hybrids, and breeders should be vigilant to developments and plants that will enable them to use apomixis in their breeding programs. Hanna and Bashaw (1987) in a recent review describe methods for identifying apomictic plants, mechanisms for using apomixis in breeding programs, and possible new developments that could expand the use of apomixis.

We believe another method that has potential for producing hybrids is the use of male gametocides. Hybrid wheat cultivars have been marketed that were produced by using male gametocides to effectively emasculate the lines used as females. It seems reasonable that some of the compounds that have been tested and proven effective as gametocides on wheat may also be effective as gametocides on forage grasses such as the wheatgrasses. Since many of these gametocides are proprietary compounds, the necessary research would have to be done in conjunction with the appropriate firms that have ownership of the compounds.

Production of Grass Hybrids—Summary

The methods that have the best potential for producing commercial grass hybrids are the use of self-incompatibility, apomixis, and gametocides. The potential increase in forage yields that could be achieved by the production of hybrid seed for use on farms definitely warrants the allocation of some breeding resources for this objective.

CELL CULTURE AND MOLECULAR GENETICS TECHNIQUES

Grass breeders to date have used conventional breeding techniques to develop new cultivars. New technologies are now becoming available for breeders to greatly expand their capabilities to solve specific breeding problems. These technologies which we will refer to as cell culture and molecular genetics techniques must be used in conjunction with conventional breeding methods because their sole use would not result in the development and use of commercial cultivars.

Tissue Culture

Techniques to culture individual plant cells and to regenerate plants from these cells have been developed for many grasses and probably can be developed for any grass by modifying the appropriate "recipes". Tissue culture gives breeders the capability to rapidly and efficiently multiply individual plants that should make commercial self-incompatible hybrids feasible if somaclonal variation can be controlled. It also enables breeders to select and apply mutagenic treatments at the cellular level (Chaleff, 1983; Schweiger et al., 1987). Cell culture can result in increased genetic variability because of somaclonal variants that can be induced by the culturing process. Cell culture permits the screening of millions of individual cells for specific traits that can be assayed at the cellular level such as resistance to specific toxins or herbicides, but it does not permit selection for many agronomically desired traits that must be investigated at the whole plant level. Mutants produced at the cellular level can be regenerated and evaluated at the whole plant level, however. Recombinant DNA work probably will be done primarily in cell culture or cell suspension systems. Techniques have been developed that permit mass culturing of cells or the culturing of individual cells (Schweiger et al., 1987). Cell culture systems and procedures will be valuable tools of present and future grass breeders.

Molecular Genetics

Molecular genetics techniques as applied to plant breeding will be used primarily to transfer traits between plants that cannot be crossed by conventional procedures (Goodman et al., 1987; Barton & Brill, 1983). The progressive complexity of the breeding procedures needed to transfer genes between organisms is shown in Table 7-6. In 1983, Barton and Brill (1983) proposed

Table 7-6. The progressive capability of gene transfer to plants by different breeding techniques. Adapted from Goodman et al. (1987).

Organism to plant transfer	Techniques used
Plant to plant, within species	Conventional breeding
Plant to plant, between species, within genera	Conventional breeding + TLC†
Plant to plant, between genera, within family or tribe	Conventional breeding + TLC + embryo rescue
Any organism to a plant	Molecular genetics + tissue culture

† TLC = Tender loving care.

that these techniques could be used to breed for insect and pest resistance, modification of seed proteins, N_2 fixation, improved photosynthesis rates, and improved stress tolerance. Goodman et al. (1987) reported that genes for herbicide and insect resistance have been transferred from bacteria into plants and that genes encoding the protein coat of tobacco mosaic virus had been inserted into tobacco (*Nicotiana tabacum* L.) and tomato (*Lycopersicum esculentum* Mill.) resulting in increased resistance to the virus. In addition to giving the breeder the capability of transferring genes from dissimilar organisms, molecular genetics techniques permit the transfer of specific genes rather than whole blocks of genes.

Specific molecular genetics approaches that offer promise for gene transfer in plants are: transfer of genes by plasmids, gene transfer by viruses, uptake or insertion of purified DNA, protoplast fusion, and *Agrobacterium*-mediated gene transfer (Goodman et al., 1987; Cocking & Davey, 1987). These and yet to be developed techniques will be used to transfer and manipulate genes in grasses. Undoubtedly, these procedures will be used in the development of grasses with tolerance to specific herbicides, insects, and diseases. We do not expect all grass breeders to become molecular geneticists, but we do consider it vital for all grass breeders to become knowledgeable in the area of molecular genetics. This knowledge will allow breeders to cooperate with molecular geneticists to solve specific problems with this new technology.

SUMMARY

There are tremendous opportunities for grass breeders to make extremely valuable contributions to humanity in the future. Grass breeders must carefully define their objectives and select those that will have the most impact on society. They must then use the most cost effective breeding methods to develop improved cultivars that meet these objectives. They also must document the added value of the results of their breeding work in economic terms and that documentation should be provided in layman's terms in order to "sell" the improved cultivars to the consuming public.

REFERENCES

Anderson, B., J.K. Ward, K.P. Vogel, M.G. Ward, H.J. Gorz, and F.A. Haskins. 1988. Forage quality and performance of yearlings grazing switchgrass strains selected for differing digestibility. J. Anim. Sci. 66:2239–2244.

Asay, K.H., J.D. Hansen, B.A. Haws, and P.O. Currie. 1983. Genetic differences in resistance of range grasses to the bluegrass billbug, *Spenophorus parvulus* (Coleptera; Curculionidae). J. Range Manage. 36:771–772.

Asay, K.H., and D.A. Johnson. 1983. Breeding for drought resistance in range grasses. Iowa State J. Res. 57:441–456.

Asay, K.H., and D.A. Johnson. 1987. Breeding for improved seedling establishment in cool-season range grasses. p. 173–176. *In* G.W. Frasier and R.A. Evans (ed.) Proc. Symp. Seed and Seedbed Ecology of Rangeland Plants, Tucson, AZ. 21–23 Apr. USDA-ARS, Natl. Tech. Infor. Serv., Springfield, VA.

Barton, K.A., and W.J. Brill. 1983. Prospects in plant genetic engineering. Science 219:671–676.

Beard, J.B. 1985. An assessment of water use by turfgrasses. p. 47–60. *In* V.A. Gilbeault and S.T. Cockerham (ed.) Turfgrass water conservation. Coop. Ext., Div. of Agriculture and Natural Resources, Univ. of California.

Braverman, S.W. 1986. Disease resistance in cool-season forage and forage range and turf grass. II. Bot Rev. 52:1–112.

Burton, G.W. 1948. The performance of various mixtures of hybrid and parent inbred millet. J. Am. Soc. Agron. 40:908–915.

Burton, G.W. 1974. Recurrent restricted phenotypic selection increases forage yields of Pensacola bahiagrass. Crop Sci. 14:831–835.

Burton, G.W. 1981. Nutrient composition of forage crops. Effects of genetic factors. p. 1–9. *In* USDA-ARS Agric. Review and Manuals. ARM-S-21. U.S. Gov. Print. Office, Washington, DC.

Burton, G.W. 1982. Improved recurrent restricted phenotypic selection increases bahiagrass forage yields. Crop Sci. 22:1058–1061.

Burton, G.W. 1985. Spaced-plant-population-progress test. Crop Sci. 25:63–65.

Burton, G.W. 1986. Developing better forages for the south. J. Anim. Sci. 63:955–961.

Campbell, W.F., B.A. Haws, K.A. Asay, and H.D. Hanse. 1984. A review of black grass bug resistance in forage grasses. J. Range Manage. 37:365–369.

Chaleff, R.S. 1983. Isolation of agronomically useful mutants from plant cell cultures. Science 219:676–682.

Chapman, H.P., W.H. Marchant, P.R. Utley, R.E. Hellwig, and W.G. Monson. 1972. Performance of steers on Pensacola bahiagrass, Coastal bermudagrass, and Coastcross-1 bermudagrass pastures and pellets. J. Anim. Sci. 34:373–378.

Clay, K., T.N. Hardy, and A.M. Hammond, Jr. 1985. Fungal endophytes of grasses and their effects on an insect herbivore. Oecologia 66:1–5.

Cockerham, S.T., and V.A. Gilbeault. 1985. The size, scope, and importance of the turfgrass industry p. 7–12. *In* V.A. Gilbeault and S.T. Cockerham (ed.) Turfgrass water conservation. Coop. Ext., Div. of Agriculture and Natural Resources, Univ. of California.

Cocking, E.G., and M.R. Davey. 1987. Gene transfer in cereals. Science 236:1259–1262.

Dewald, C.L., and R.S. Dayton. 1985. Registration of gynomonoecious germplasm (GSF-I and GSF-II) of eastern gamagrass. Crop Sci. 25:715.

Empig, L.T., C.O. Gardner, and W.A. Compton. 1972. Theoretical gains for different population improvement procedures. Nebraska Agric. Exp. Stn. MP 26.

Falconer, D.S. 1981. Introduction to quantitative genetics. 2nd ed. Longman, New York.

Flanery, J.A. 1987. Thirsty Colorado goes after South Platte river. Sunday Omaha World Herald, Omaha, NE. 1 Nov. p. 1, 12.

Funk, C.R., P.M. Halisky, M.C. Johnson, M.R. Siegel, A.V. Stewart, S. Ahmad, R.H. Hurley, and I.C. Harvey. 1983. An endophytic fungus and resistance to sod webworms: association in *Lolium perenne* L. Bio/Technology 1:189–191.

Glewen, K.L., and K.P. Vogel. 1984. Partitioning the genetic variability for seedling growth in sand bluestem into its seed size and seedling vigor components. Crop Sci. 24:137–141.

Goodman, R.M., H. Hauptl, A. Crossway, and V. Knauff. 1987. Gene transfer in crop improvement. Science 236:48–54.

Hallauer, A.R., and J.B. Miranda, Fo. 1981. Quantitative genetics in maize breeding. Iowa State Univ. Press, Ames.

Hanna, W.W., and E.C. Bashaw. 1987. Apomixis: Its identification and use in plant breeding. Crop Sci. 27:1136–1139.

Hanson, A.A., and H.L. Carnahan. 1956. Breeding perennial forage grasses. Tech. Bull. 1145. USDA-ARS, U.S. Gov. Print. Office, Washington, DC.

Hoveland, C.J., S.P. Schmidt, C.C. King Jr., J.W. Odom, E.M. Clark, J.A. McGuire. L.A. Smith, H.W. Grimes, and J.L. Holliman. 1983. Steer performance and association of *Acremonium coenophialum* fungal endophyte on tall fescue pasture. Agron. J. 75:821–824.

Johnson, M.C., D.L. Dahlman, M.R. Siegel, L.P. Bush, G.C.M. Latch, D.A. Potter, and D.R. Varney. 1985. Insect feeding deterrents in endophyte-infected tall fescue. Applied and Environ. Microbiol. 49:568–571.

Kube, J.G., K.P. Vogel, and L.E. Moser. 1989. Genetic variability for seedling atrazine tolerance in Indiangrass (*Sorghastrum nutans*). Crop Sci. 29:18–23.

Marten, G.C., R.M. Jordan, and A.W. Hovin. 1976. Biological significance of reed canarygrass alkaloids and associated palatability variation to grazing sheep and cattle. Agron. J. 68:909–914.

Martin, A.R., R.S. Moomaw, and K.P. Vogel. 1982. Warm-season grass establishment with atrazine. Agron. J. 74:916–920.

McKell, C.M. 1972. Seedling vigor and seedling establishment. p. 76–89. *In* V.B. Youngner and C.M. McKell (ed.) The biology and utilization of grasses. Academic Press, New York.

Nelson, C.J., D.A. Sleper, and J.H. Coutts. 1985. Field performance of tall fescue selected for leaf-area expansion. p. 320–322. *In* Proc. Int. Grassl. Congr. Kyoto, Japan. 24–31 Aug. The Sci. Counc. Japan and the Japanese Soc. Grassl., Nishi-nasuno, Tochigi-ken, Japan.

Nguyen, H.T., and D.A. Sleper. 1983. Theory and applications of half-sib matings in forage grass breeding. Theor. Appl. Genet. 64:187–196.

Schweiger, H.G., J. Dirk, H.-U. Koop, E. Kranz, G. Neuhaus, G. Spangenberg, and D. Wolff. 1987. Individual selection, culture, and manipulation of higher plant cells. Theor. Appl. Genet. 73:769–783.

Shearman, R.C. 1986. Kentucky Bluegrass cultivar evapotranspiration rates. HortSci. 21:455–457.

Shearman, R.C., and J.B. Beard. 1975. Turfgrass wear tolerance mechanisms. II. Effects of cell wall constituents on turfgrass wear tolerance. Agron. J. 211–215.

Stimman, M.W., and C.M. Taliaferro. 1969. Resistance of selected accessions of bermudagrass to phytotoxemia caused by adult two-lined spittlebugs. J. Econ. Entomol. 62:1189–1190.

Vogel, K.P. 1987. Seeding rates for establishing big bluestem and switchgrass with pre-emergence atrazine applications. Agron. J. 79:509–512.

Vogel, K.P., R. Britton, H.J. Gorz,and F.A. Haskins. 1984. In vitro and in vivo analyses of hays of switchgrass strains selected for high and low in vitro dry matter digestibility. Crop Sci. 24:977–980.

Vogel, K.P., F.A. Haskins, and H.J. Gorz. 1981. Divergent selection for in vitro dry matter digestibility in switchgrass. Crop Sci. 21:39–41.

Voight, P.W., C.R. Tischler, and B.A. Young. 1987. Selection for improved establishment in warm-season grasses. p. 177–187. *In* G.W. Frasier and R.A. Evans (ed.) Proc. Symp., Seed and Seedbed Ecology of Rangeland Plants, Tucson, AZ. 21–23 Apr. USDA-ARS, Natl. Tech. Inf. Serv., Springfield, VA.